INTRODUCTION TO STRUCTURES

D1333852

INTRODUCTION TO STRUCTURES

W. R. SPILLERS, B.Sc., M.Sc., Ph.D.
Professor of Civil Engineering
Rensselaer Polytechnic Institute, Troy, New York

ELLIS HORWOOD LIMITED
Publishers · Chichester

Halsted Press: a division of
JOHN WILEY & SONS
New York · Chichester · Brisbane · Toronto

First published in 1985 by

ELLIS HORWOOD LIMITED
Market Cross House, Cooper Street, Chichester, West Sussex, PO19 1EB, England

The publisher's colophon is reproduced from James Gillison's drawing of the ancient Market Cross, Chichester.

Distributors:

Australia, New Zealand, South-east Asia:
Jacaranda-Wiley Ltd., Jacaranda Press,
JOHN WILEY & SONS INC.,
G.P.O. Box 859, Brisbane, Queensland 40001, Australia

Canada:
JOHN WILEY & SONS CANADA LIMITED
22 Worcester Road, Rexdale, Ontario, Canada.

Europe, Africa:
JOHN WILEY & SONS LIMITED
Baffins Lane, Chichester, West Sussex, England.

North and South America and the rest of the world:
Halsted Press: a division of
JOHN WILEY & SONS
605 Third Avenue, New York, N.Y. 10016, U.S.A.

© 1985 W.R. Spillers/Ellis Horwood Limited

British Library Cataloguing in Publication Data
Spillers, W.R.
Introduction to structures. —
(Ellis Horwood series in engineering science)
1. Structural engineering
I. Title
624 TA633

Library of Congress Card No. 84-12963

ISBN 0-85312-725-5 (Ellis Horwood Limited — Library Edn.)
ISBN 0-85312-558-9 (Ellis Horwood Limited — Student Edn.)
ISBN 0-470-20094-4 (Halsted Press)

Printed in Great Britain by R.J. Acfords, Chichester.

Contents

Contents

Contents

Foreword

Over the past 20 years massive changes have occurred in the practice of structural analysis. Those of us who graduated from an engineering school in the 1950s were taught relatively simple skills and hoped that we would never be asked to analyse structures with more than three redundants. Today a structure with 300 redundants is not considered large or thought to present any particular problem for analysis. In fact, a highly reliable solution for such a structure is usually available at a modest price. We have thus gone from a situation in which the analysis of a highly indeterminate structure posed considerable difficulty to one in which analysis capabilities are readily available and cheap.

Educational institutions have for the most part been slow to respond to these changes. While we did institute computer programming courses quickly and in some cases graduate courses in computer-aided structural analysis, surveys have shown that the undergraduate curriculum – the backbone of professional engineering – is roughly the same as it was 20 to 30 years ago. To the extent that we teach engineering, not computer programming, this lack of response to the advent of the computer may have been appropriate. But in the long run the computer will surely have its impact upon the way we teach structures. While not presuming to know how matters will eventually turn out, it is the thesis of this text that because of the computer (if for no other reason) structural engineers must know more today.

The question is, of course, how to know more. As far as this text is concerned that question is answered in two ways. In terms of depth, an attempt has been made to discuss three-dimensional problems more than has been common in the past. In terms of scope, the text moves through statically indeterminate structures and on some plastic analysis. In order to do this it has been necessary to omit some (in this context) redundant topics such as the conjugate beam and the three-moment equations.

Otherwise, the outline of this text is straightforward. It moves logically from statically determinate structures to the computation of displacements to the analysis of statistically indeterminate structures. Then follow four supplementary chapters dealing with plastic analysis, cables, moment distribution, and influence lines. In terms of style, there is a tendency to include more material than the reader might want on first reading. That

Foreword

is done in the hope that he or she will return for a second look and even try the references which are indicated.

There is a 200-year tradition in structures. As a result, those of us who call ourselves structural engineers spend much time learning – really taking – from others. In my own case this includes not only teachers and colleagues but also long-suffering students at both Columbia University and Rensselaer Polytechnic Institute to whom I am grateful. What we have shared is a common interest in how structures work.

Finally a word about Ellis Horwood and his publishing house. At a time of harsh ecconomic realities, Ellis Horwood turns out to be a creative man of great energy and enthusiasm. He is not simply a man of his word; he returns publishing to what we commonly think of as better times.

William R. Spillers

1. Introduction and Review

1.1 MODELING

Structural analysis, the subject of this text, is for the most part concerned with finding the structural response (the lateral deflection of a building under wind load, the reaction of a bridge to a moving train . . .) given the external loads. In all but the most trivial cases, real structures, that is structures without the simplifications commonly associated with analysis, turn out to be impossibly complex. And what is finally analyzed – the structural model – may appear at first glance to be quite different than the real structure.

Constructing a structural model of a given physical situation involves discarding certain features and emphasizing others in an attempt to develop a 'reasonable' representation. In doing so the engineer must exercise judgement in knowing what to discard and when he has reached a workable model. This brings up the difference between engineering and analysis.

This text is concerned with analysis, not engineering. Given the structural model and the type of analysis to be performed, actually performing the analysis should be a matter of routine and not involve engineering judgement. However, even with analysis, engineering judgement is required at two points. It is first of all necessary to use engineering judgement to construct the model, given the real structure. At some later point in time, given the analysis, the engineer must use judgement to decide – for whatever reasons – whether or not his results make sense.

It is not possible to over-emphasize the importance of these two steps. Eventually the engineer must 'accept' his analysis and move forward with the process of design and construction. If an error of analysis leads to a design failure, he cannot simply shrug his shoulders and walk off. He is legally and ethically responsible for producing a design which functions adequately. In practical terms the only way this can be done is through developing an 'understanding' of his structure to the extent that he knows how the analysis will turn out before he actually does it. The curious part is that this understanding is developed through performing analyses and thus one of the facets of this text.

Modeling may proceed on many levels:

(1) *Structural modeling.* Elementary structural analysis is concerned with *skeletal structures* or structures which can be represented by lines and properties associated with lines. For example, the primary analysis of the Rio–Niterói Bridge (see the frontispiece) was probably performed on a structural model which was a beam, represented by a single line.

   In order to learn to model structures properly, it is important for the engineer to observe structures and try to understand how each structure functions. (This is equivalent to making a structural model in your mind as you pass a structure.) When the functioning of a structure is obvious, so is its structural model. The truss bridge schematic of Fig. 1.1 is a case in point. Here the primary structural elements are the

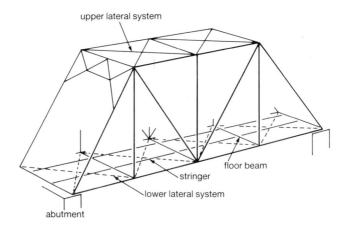

Fig. 1.1 – Schematic or 'model' of a truss bridge.

parallel trusses which transfer the loads from the bridge itself to the abutments. The typical load path involves a load on the bridge deck which is transferred to the stringers which are supported by the floor beams which frame into the truss joints. The upper and lower lateral systems are concerned with lateral load (e.g. wind) transfer and bracing against buckling.

   A similar analysis can be made of the industrial building of Fig. 1.2. Schematically, the roof loads are transferred by the purlins to the roof trusses which are supported by columns. The bracing systems again are primarily concerned with lateral load and buckling.

There is a full spectrum of structural complexity. While the two structures just mentioned function in rather obvious manners, a point

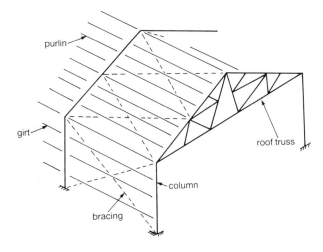

Fig. 1.2 – Schematic or 'model' of an industrial building.

load applied to a spherical shell or even a cable net (see Chater 6) can produce a complex set of reactions which can be difficult to anticipate. In these latter cases it is even more important that the engineer develop some way in which he can 'understand' the structure for which he is responsible.

(2) *Envionmental modeling.* The term environmental modeling is first of all used here in connection with loads. While the dead load or weight of a structure should by definition be known accurately to its designer, most other cases turn out to be less clearly defined. The engineer frequently does not know precisely the uses to which his building will be put during its lifetime, he cannot anticipate all possible combinations of cars and trucks which will use his bridge, some loads such as wind, show, and earthquakes possess a high degree of randomness . . .

From a practical point of view, many of these questions have been studied carefully over the years and in many cases the engineer *must* use loads specified in various *codes* of practice. For example, the Building Officials and Code Administrators (BOCA) code specifies various uniform live loads (loads associated with type of usage) for buildings as indicated in Table 1.1. Similarly, the American Association of State Highway and Transportation Officials

Table 1.1 Typical live loads for buildings

THE BOCA BASIC BUILDING CODE/1978

MINIMUM UNIFORMLY DISTRIBUTED LIVE LOADS

Occupancy or use	Live load (psf)
Apartments (see Residential)	
Armories and drill rooms	150
Assembly halls and other places of assembly:	
Fixed seats	60
Movable seats	100
Platforms (assembly)	100
Balcony (exterior)	100
One- and two-family dwellings only	60
Bowling alleys, poolrooms, and similar recreational areas	75
Cornices	75
Court rooms	100
Corridors	100
First floor	100
Other floors, same as occupancy served except as indicated	
Dance halls and ballrooms	100
Dining rooms and restaurants	100
Dwellings (see Residential)	100
Fire escapes	100
On multi- or single-family residential buildings only	40
Garages (passenger cars only)	50
For trucks and buses use AASHO[1] lane loads	
(see Table 707 for concentrated load requirements)	
Grandstands (see Reviewing stands)	
Gymnasiums, main floors and balconies	100
Hospitals	
Operating rooms, laboratories	60
Private rooms	40
Wards	40
Corridors, above first floor	80
Hotels (see Residential)	
Libraries:	
Reading rooms	60
Stack rooms (boc) & shelving at 65 pcf) but not less than	150
Corridors, above first floor	80
Manufacturing:	
Light	125
Heavy	250
Marquees	75
Office buildings:	
Offices	50
Lobbies	100
Corridors, above first floor	80
File and computer rooms require heavier loads based upon	
anticipated occupancy	80

Open parking structures (passenger cars only)	50
Penal institutions:	
Cell blocks	40
Corridors	100
Residential:	
Multifamily houses	
Private apartments	40
Public rooms	100
Corridors	80
Dwellings:	
First floor	40
Second floor and habitable attics	30
Uninhabitable attics[2]	20
Hotels:	
Guest rooms	40
Public rooms	100
Corridors serving public rooms	100
Corrisors	80
Reviewing stands and bleachers[3]	100
Schools:	
Classrooms	40
Corridors	80
Sidewalks, vehicular driveways, and yards subject to trucking	250
Skating rinks	100
Stairs and exitways	100
Storage warehouse:	
Light	125
Heavy	250
Stores:	
Retail:	
First floor, rooms	100
Upper floors	75
Wholesale	125
Theatres:	
Aisles, corrisors, and lobbies	100
Orchestra floors	60
Balconies	60
Stage floors	150
Yards and terraces, pedestrians	100

Note 1. American Association of State Highway Officials.

Note 2. Live load need be applied to joists or to bottom chords of trusses or trussed rafters only in those portions of attic space having a clear height of forty-two (42) inches or more between joist and rafters in conventional rafter construction, and between bottom chords and any other member in trusses or trussed rafters shall be designed to sustain the imposed dead load or ten pounds per square foot (10 psf) whichever be greater, uniformly distributed over the entire pan.

A further ceiling dead load reduction to a minimum of five pounds per square foot (5 psf) or the actual dead load, whichever is greater, may be applied to joists in conventional rafter construction or to the bottom chords of trusses or trussed rafters under either or both of the following conditions:

    a. If the clear height is not over thirty (30) inches between joists and rafters in conventional construction and between the botton chord and any other member for trusses or trussed rafter construction.

    b. If a clear height of greater than thirty (30) inches as defined in 'a' directly above, does not exist for a horizontal distance of more than twelve (12) inches along the member.

Note 3. For detailed recommendations see The Standard for Tents, Grandstands, and Air-Supported Structures Used for Places of Assembly listed in Appendix B.

(AASHTO) code specifies standard trucks (see Fig. 1.3) to be used in the design of highway bridges, and the American Railway Engineering Association (AREA) specifies standard trains for railway bridges (see Fig. 1.4).

By definition, the design of a 'conventional' structure must follow some given building code. In the interesting case of a new type of structure or a structure with monumental proportions, the engineer will go to great lengths to ensure that his design is adequate. For example, wind loading on a fabric structure of unusual shape and certainly wind loading on a building of record height commonly require model studies in wind tunnels (at considerable expense). Finally, there are common problems such as foundation settlement which routinely require soil samples taken in the field to be tested in the laboratory. Less common are special problems of a corrosive environment (certain types of manufacturing, sanitary sewers . . .), dynamic effects (heavy manufacturing, crane loads . . .), wave action on ocean platforms . . . The list is endless but the point to be made is clear. The engineer must first understand the environment of his structure and then design for it.

(3) *Material modeling.* The treatment of structural materials is another area in which it is common to make engineering approximations within a structural analysis. So far as this text is concerned, two types of material assumptions will be made. A material will be assumed to be

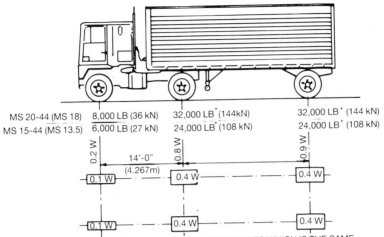

MS 20-44 (MS 18)   8,000 LB (36 kN)   32,000 LB* (144kN)   32,000 LB* (144 kN)
MS 15-44 (MS 13.5) 6,000 LB (27 kN)   24,000 LB* (108 kN)   24,000 LB* (108 kN)

W = COMBINED WEIGHT ON THE FIRST TWO AXLES WHICH IS THE SAME
     AS FOR THE CORRESPONDING H (M) TRUCK
V = VARIABLE SPACING-14 FT TO 30 FT (4.267 to 9.144mm) INCLUSIVE
     SPACING TO BE USED IS THAT WHICH PRODUCES MAXIMUM STRESSES

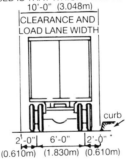

STANDARD HS (MS) TRUCKS

In the design of timber floors and orthotropic steel decks (excluding transverse beams) for H20 (M 18) loading, one axle load of 24 000 lb (108 kN) or two axle loads of 16 000 lb (72 kN) each, spaced 4 ft (1.219 m) apart may be used, whichever produces the greater stress, instead of the 32 000 lb (144 kN) axle shown.

For slab design, the center line of wheels shall be assumed to be 1 ft (0.305 m) from face of curb. (See Art. 1.3.2(B))

Fig. 1.3 – Examples of standard trucks used in bridge design. (Reprinted with permission of the American Association of State Highway and Transportation Officials.)

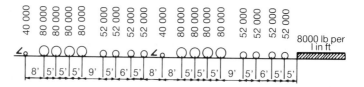

Fig. 1.4 – Example of a standard RR bridge loading. (Reprinted with permission of the American Railway Engineering Association.)

either elastic or elastic-plastic (somes rigid-plastic). Chapter 5 will return briefly to the question of real material properties; only some simple points will be noted here:

(a) No real structural material is perfectly elastic.
(b) Materials such as wood, concrete, and steel (at high temperatures) have significant time-dependent material responses.
(c) Some materials such as concrete and aluminum exhibit nonlinear behavior under 'working' loads.
(d) Some structural materials such as some types of plastics degrade with time and exposure to sunlight indepdent of their load level. This can be important in the design of fabric roofs.
(e) Problems of corrosion and fatigue can be significant in structural design situations.
(f) While it is normally assumed that unloaded structural elements

are stress free, manufacturing processes can in some cases be quite detrimental. Typical examples of this include stresses induced through the 'rolling' of wide flange sections, punching for rivets and bolts, and welding which is known to crack thick steel plates when not done properly.

(4) *Linear versus non-linear analysis.* Without comment beyond this section, this text is concerned with linear analysis in which the equilibrium equations of an element are written in its undeformed configuration. (A comparable approximation is made when length change is computed from member displacements.) Put more simply, while in real life structures are in equilibrium in their deformed positions, it is certainly easier and in many cases adequate to neglect these deformations. Doing so results in linear equations which must be solved (linear analysis). Including member deformations generally leads to nonlinear analysis which is of course more complex to perform. While linear analysis is adequate in the most commonly encountered situations, there are two situations where it does *not* work: buckling and some problems of cable nets.

The point of this discussion of modeling is to raise issues rather than solve them. This book is not, in fact, concerned directly with questions of modeling. As a rule, certain classes of structures and their loads will be given for analysis. If a given structure can be approximated by one which contains bars which have only axial stiffness (springs in classical mechanics) then the structure is called a *truss*; if there is bending present within structural members, the structure is called a *frame*; if the structure is composed of flat, sheet-like elements rather than line elements, it is called a *plate*; if these sheet-like elements are curved rather than flat, then the structure is called a *shell*. There can of course be hybrid structures but these can at worse be subsumed within the more complex class of structures involved.

1.2 NOTATION

This section is concerned with vector notation and the rules of vector algebra. It is assumed that the reader is already familiar with the concept of force, moment, displacement, and (small) rotation as vector quantities from elementary physics and mechanics courses. For that reason a fairly rapid pace will be maintained; applications will be stressed.

From the point of view of this text, vector notation is simply a short-hand and a vector equation, a simple means of representing two scalar equations when dealing with plane problems or three scalar equations when dealing with space problems. The following discussion will center on the three-dimensional case; the plane case will be relegated to examples.

A vector **A** is described by listing its three components. That is commonly done in three different ways:

(a) *ijk* notation

$$\mathbf{A} = A_x\mathbf{i} + A_y\mathbf{j} + A_z\mathbf{k} \tag{1.1}$$

Here **i**, **j**, **k** are the usual coordinate unit vectors or base vectors and A_x, A_y, A_z are called the components of **A**.

(b) Ordered triplet notation

$$\mathbf{A} = (A_x, A_y, A_z) \tag{1.2}$$

(c) Matrix notation

$$A = \begin{bmatrix} A_x \\ A_y \\ A_z \end{bmatrix} \tag{1.3}$$

These notations all have their uses but for the most part the *ijk* notation is regarded in this text to be unnecessarily cumbersome.

Some of the common vector definitions and operations are:

Zero vector   A vector is said to be zero when each of its components is zero. That is

$$\mathbf{A} = 0 \quad \Rightarrow \quad A_x = A_y = A_z = 0$$

Vector equality   Two vectors are said to be equal when their components are equal. That is

$$\mathbf{A} = \mathbf{B} \quad \Rightarrow \quad \begin{array}{l} A_x = B_x \\ A_y = B_y \\ A_z = B_z \end{array} \qquad (1.4)$$

Magnitude of a vector   In geometric terms, if a vector $\mathbf{A}$ is represented by a line segment from the origin to a point whose coordinates are the vector components, the magnitude of $\mathbf{A}$, written $|\mathbf{A}|$, is simply the length of the line from the origin to the point. That is

$$|\mathbf{A}| = (A_x^2 + A_y^2 + A_z^2)^{1/2} \qquad (1.5)$$

Unit vector   A unit vector $\mathbf{n}$ is defined to have unit magnitude. That is

$$|\mathbf{n}| = 1 \qquad (1.6)$$

Vector addition

$$\mathbf{A} + \mathbf{B} = \mathbf{C} \quad \Rightarrow \quad \begin{array}{l} A_x + B_x = C_x \\ A_y + B_y = C_y \\ A_z + B_z = C_z \end{array} \qquad (1.7)$$

Scalar product

$$\mathbf{A} \cdot \mathbf{B} = C \quad \Rightarrow \quad C = A_x B_x + A_y B_y + A_z B_z \qquad (1.8)$$

(Note that $|\mathbf{A}| = (\mathbf{A} \cdot \mathbf{A})^{1/2}$)

Vector product

$$\mathbf{A} \times \mathbf{B} = \mathbf{C} \quad \Rightarrow \quad \begin{array}{l} C_x = A_y B_z - A_z B_y \\ C_y = A_z B_x - A_x B_z \\ C_z = A_x B_y - A_y B_x \end{array} \qquad (1.9)$$

Projection or component   The projection A_n of vector $\mathbf{A}$ in a direction defined by a unit vector $\mathbf{n}$ is defined to be

$$A_n = \mathbf{A} \cdot \mathbf{n} \qquad (1.10)$$

In mechanics it is common to use a unit vector to represent the slope (or direction) of a line. For example, given two points A and C and their position vectors $\mathbf{R}_A$ and $\mathbf{R}_C$, a unit vector pointing in the direction from C to A can be written as

$$\mathbf{n} = \frac{\mathbf{R}_A - \mathbf{R}_C}{|\mathbf{R}_A - \mathbf{R}_C|} \qquad (1.11)$$

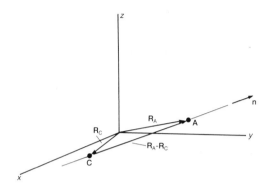

Fig. 1.5 – Construction of a unit vector **n** along a line from point C to point A.

This expression follows directly from Fig. 1.5 since the vector $\mathbf{R}_A - \mathbf{R}_C$ has the desired direction and dividing a vector by its magnitude produces a unit vector. Example 1.1 shows the calculations that are required to compute a typical unit vector.

Example 1.1   Given points A and C, construct a unit vector **n** along a line from C to A. Given:

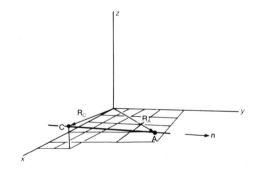

$$\mathbf{R}_A = (4, 4, 1)$$
$$\mathbf{R}_C = (5, 2, 2)$$

Step 1. Compute the components of $\mathbf{R}_A - \mathbf{R}_C$
$$\mathbf{R}_A - \mathbf{R}_C = (-1, 2, -1)$$

Step 2. Compute the magnitude of $\mathbf{R}_A - \mathbf{R}_C$

$$|\mathbf{R}_A - \mathbf{R}_C| = \sqrt{6}$$

Step 3. Divide $\mathbf{R}_A - \mathbf{R}_C$ by its magnitude to obtain the unit vector $\mathbf{n}$

$$\mathbf{n} = (-1, 2, -1)/\sqrt{6} = (-0.408, 0.816, -0.408)$$

1.3 RIGID BODY MECHANICS

This section will be used to review some of the fundamental concepts of mechanics and indicate some immediate applications to structures. First of all there is the concept of the (vector) moment of a force $\mathbf{F}$ about any point O (see Fig. 1.6). Given $\mathbf{F}$ and the position vector $\mathbf{r}$ which describes

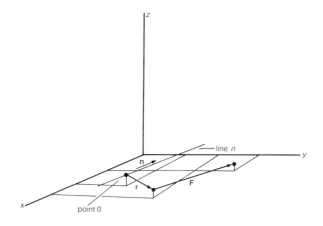

Fig. 1.6 – Vector moment of force $\mathbf{F}$ about point 0.

its location with respect to point O, the moment of $\mathbf{F}$ about point O is defined to be

$$\mathbf{M}_0 = \mathbf{r} \times \mathbf{F} \tag{1.12}$$

The (scalar) moment M_n of $\mathbf{F}$ about a line n (whose direction is defined by the unit vector $\mathbf{n}$) through point O appears as the derived quantity

$$M_n = \mathbf{n} \cdot \mathbf{M}_0 \tag{1.13}$$

Equations (1.12) and (1.13) are simply compact descriptions of familiar physical quantities. For example, the components of $\mathbf{M}_0$ are simply the quantities ordinarily associated with the scalar moments of the force **F** about lines through point O parallel to coordinate axes. This can be seen by writing out these components

$$(\mathbf{M}_0)_x = r_y F_z - r_z F_y$$
$$(\mathbf{M}_0)_y = r_z F_x - r_x F_z \qquad\qquad (1.14)$$
$$(\mathbf{M}_0)_z = r_x F_y - r_y F_x$$

The x component of $\mathbf{M}_0$ (see Fig. 1.7) is clearly the moment of **F** about a

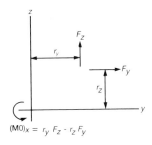

$$(M0)_x = r_y F_z - r_z F_y$$

Fig. 1.7 – x component of $\mathbf{M}_0$ shown using the right-hand rule.

line through O parallel to the x-axis. (Similar comments are valid for the other components.)

The scalar moment M_n describes the tendency of **F** to twist a shaft along the line defined by **n** (see Fig. 1.8). That fact can be seen by writing

$$M_n = \mathbf{n} \cdot (\mathbf{r} \times \mathbf{F}) = (\mathbf{n} \times \mathbf{r}) \cdot \mathbf{F} \qquad\qquad (1.15)$$

and noting that the term $(\mathbf{n} \times \mathbf{r})$ is perpendicular to both **n** and **r** and has the magnitude R of the shaft in Fig. 1.8. The scalar product of the force

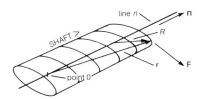

Fig. 1.8 – Scalar moment of force **F** about line n.

F with this term then becomes the product of the component of **F** normal to the shaft times the radius R as suggested above.

Given a rigid body under a small rotation about an axis (see Fig. 1.9) the displacement **δ** is given by

$$\boldsymbol{\delta} = \boldsymbol{\omega} \times \mathbf{r} \tag{1.16}$$

Here $\boldsymbol{\delta}$ is the displacement or change in position of a point located by the position vector $\mathbf{r}$, $\boldsymbol{\omega}$ is the (small) rotation vector, and $\mathbf{r}$ is the position vector of any point in the body with respect to a point on the axis of rotation.

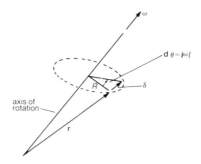

Fig. 1.9 – Small rigid body rotation about an axis.

Somewhat heuristically, let the body rotate about an axis (whose direction is defined by the vector $\boldsymbol{\omega}$) through some small angle $d\theta = |\boldsymbol{\omega}|$. The motion of any point is then normal to both $\boldsymbol{\omega}$ and $\mathbf{r}$ and has the magnitude $R\, d\theta$. That is precisely the motion described by Eq. (1.16). It follows that an arbitrary (small) rigid body displacement $\boldsymbol{\delta}$ can be written as the sum of a (constant) translation $\boldsymbol{\delta}_0$ plus a rotation term given by Eq. (1.16) or

$$\boldsymbol{\delta} = \boldsymbol{\delta}_0 + \boldsymbol{\omega} \times \mathbf{r} \tag{1.17}$$

In terms of this notation, it is now possible to write the equations of equilibrium for any rigid body (see Fig. 1.10). In this case the rigid body is acted upon by n forces $\mathbf{F}_i$ and m moments $\mathbf{M}_i$. The equations of equilibrium are simply:

(1) The forces must sum to zero

$$\sum_{i=1}^{n} \mathbf{F}_i = 0 \tag{1.18}$$

(2) The moments must sum to zero about any point O

$$\sum_{i=1}^{n} \mathbf{r}_i \times \mathbf{F}_i + \sum_{j=1}^{m} \mathbf{M}_j = 0 \tag{1.19}$$

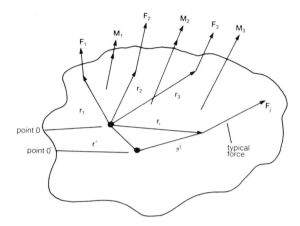

Fig. 1.10 – Typical rigid body acted on by forces $\mathbf{F}_i$ and moments $\mathbf{M}_i$.

It follows directly that:

THEOREM 1.1 *If the forces sum to zero and the moments sum to zero about some point O, then the moments about any point O' must also sum to zero.*

Proof Let ρ_i represent the position vectors which locate the forces $\mathbf{F}_i$ with respect to the new point O' and $\mathbf{r}'$ the translation of the origin (see Fig. 1.10). It follows that $\mathbf{r}_i = \rho_i - \mathbf{r}'$. Using this transformation the theorem follows directly as

$$\sum_{i=1}^{n} \mathbf{r}_i \times \mathbf{F}_i + \sum_{j=1}^{m} \mathbf{M}_j = 0$$

$$\sum_{i=1}^{n} (\rho_i - \mathbf{r}') \times \mathbf{F}_i + \sum_{j=1}^{m} \mathbf{M}_j = 0$$

$$\sum_{i=1}^{n} \rho_i \times \mathbf{F}_i - \mathbf{r}' \times \sum_{i=1}^{n} \mathbf{F}_i + \sum_{j=1}^{m} \mathbf{M}_j = 0$$

$$\sum_{i=1}^{n} \rho_i \times \mathbf{F}_i + \sum_{j=1}^{m} \mathbf{M}_j = 0$$

Equations (1.18) and (1.19) are the very basic equations of equilibrium of a rigid body. They state that the forces must sum to zero in three directions (the x, y, z coordinate directions) for three-dimensional problems and in two directions in plane problems. The moment equations are somewhat more difficult. They imply that the moments must sum to zero about the three coordinate axes in space and about a single axis (out of the plane) in two-dimensional problems. Theorem 1.1 is conceptually useful. It argues that once the three equations of moment equilibrium have been used, it is not possible to obtain additional information (independent equations) by taking moments about another point of the same rigid body.

In terms of these variables, the principle of virtual work for a rigid body follows directly:

THEOREM 1.2 *The virtual work of a rigid body in equilibrium under a small rigid body motion is zero.*

Proof

$$\sum_{i=1}^{n} \mathbf{F}_i \cdot \boldsymbol{\delta}_i + \sum_{j=1}^{m} \mathbf{M}_j \cdot \boldsymbol{\omega} = \sum_{i=1}^{n} \mathbf{F}_i \cdot (\boldsymbol{\delta}_0 + \boldsymbol{\omega} \times \mathbf{r}_i) + \sum_{j=1}^{m} \boldsymbol{\omega} \cdot \mathbf{M}_j$$

$$= \boldsymbol{\delta}_0 \cdot \left(\sum_{i=1}^{n} \mathbf{F}_i \right) + \boldsymbol{\omega} \cdot \left(\sum_{j=1}^{m} \mathbf{M}_j + \sum_{i=1}^{n} \mathbf{r}_i \times \mathbf{F}_i \right) = 0$$

In this theorem, the left-hand side is simply the sum of the work done by the forces moving through their respective displacements and the moments under the constant rotation; the right-hand side is shown to be zero, using the equilibrium equations, Eqs (1.18) and (1.19).

1.3.1 The Tripod
Example 1.2 shows an application of the equations of statics to the classic problem of computing the reactions of a three-bar truss, given the load.

Example 1.2 Given the truss shown below, find the reactions F_1, F_2, F_3.

Step 1. Compute unit vector components
Step 2. Write equilibrium equations
Step 3. Solve for F_1, F_2, F_3

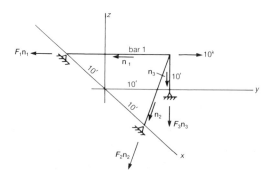

Unit vectors:

$$\mathbf{n}_1 = (-1/\sqrt{3}, -1/\sqrt{3}, -1/\sqrt{3})$$

$$\mathbf{n}_2 = (1/\sqrt{3}, -1/\sqrt{3}, -1/\sqrt{3})$$
$$n_3 = (0, 0, -1)$$

Equilibrium equations:

$$-F_1/\sqrt{3} + F_2/\sqrt{3} \qquad\qquad = 0$$

$$-F_1/\sqrt{3} - F_2/\sqrt{3} \qquad + 10 = 0$$

$$-F_1/\sqrt{3} - F_2/\sqrt{3} - F_3 \qquad = 0$$

Add first two equations $\quad \Rightarrow \quad -\dfrac{2}{\sqrt{3}} F_1 + 10 = 0 \qquad F_1 = \dfrac{10\sqrt{3}^k}{2}$

Subtract first two equations $\Rightarrow \dfrac{2}{\sqrt{3}} F_2 - 10 = 0 \qquad F_2 = \dfrac{10\sqrt{3}^k}{2}$

Third equation $\qquad\qquad\qquad \Rightarrow \qquad\qquad\qquad F_3 = -10^k$

Note that moments about the x-axis gives F_3 directly.

$$F_3 \times 10 + 10 \times 10 = 0 \quad \Rightarrow \quad F_3 = -10^k$$

It is necessary, first of all, to assume that these reactions are parallel to the bars themselves and can be written using unit vectors as indicated in the example. This point will be argued more carefully later.

The most obvious way to proceed is the write the equations of force equilibrium as

$$F_1\mathbf{n}_1 + F_2\mathbf{n}_2 + F_3\mathbf{n}_3 + \mathbf{P} = 0 \qquad \text{(vector form)}$$

or

$$F_1(\mathbf{n}_1)_x + F_2(\mathbf{n}_2)_x + F_3(\mathbf{n}_3)_x + (\mathbf{P})_x = 0$$
$$F_1(\mathbf{n}_1)_y + F_2(\mathbf{n}_2)_y + F_3(\mathbf{n}_3)_y + (\mathbf{P})_y = 0 \qquad \text{(scalar form)}$$
$$F_1(\mathbf{n}_1)_z + F_2(\mathbf{n}_2)_z + F_3(\mathbf{n}_3)_z + (\mathbf{P})_z = 0$$

Example 1.2 shows in detail how these equations can be solved in a specific case. Incidentally, this example also shows that the moment equation about the x-axis can be used directly to compute F_3. This example will be returned to in the next chapter.

1.3.2 The Two-Bar (Plane) Truss

Example 1.3 shows the plane analog of the three-bar truss just discussed. Again note that rather than solving the simultaneous equations of equilibrium, a moment equation can be used to compute each reaction directly.

Example 1.3   Find the reactions F_1 and F_2 for the poane truss shown.

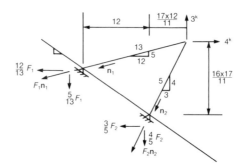

Equilibrium equations:

$$\frac{12}{13}F_1 + \frac{3}{5}F_2 = 4 \qquad F_1 = \frac{7 \times 13^k}{33}$$

$$\Rightarrow$$

$$\frac{5}{13}F_1 + \frac{4}{5}F_2 = 3 \qquad F_2 = \frac{80^k}{33}$$

Alternative: Compute F_2 directly by taking moments about the upper support.

$$\frac{3}{5}F_2 \times 12 + \frac{4}{5}F_2 \times 12 = 3 \times \left(12 + \frac{17 \times 12}{11}\right) - 4 \times \left(\frac{16 \times 17}{11} - 12\right)$$

$$F_2 = \frac{80^k}{33} \text{ (again)}$$

1.4 THE CONCEPT OF STRESS RESULTANTS

As indicated above, one of the primary goals of this text is the computation of internal member forces given the external loads. This is done by 'cutting' a structure at the point of interest (see Fig. 1.11) and then attempting to find the right combination of free body diagrams to allow the internal member forces to be computed using the equations of equilibrium. These internal member forces which appear when a structure is cut are called

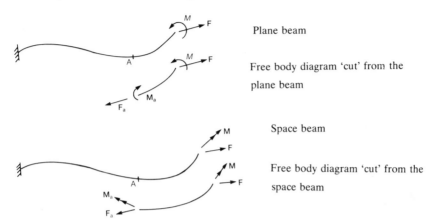

Note that internal forces must appear on the free body diagram of a 'cut' structure.

Fig. 1.11 – Stress resultants.

stress resultants because they are statically equivalent to the stresses distributed over the cross-section which has been cut.

There are two points to be made in connection with stress resultants here in this section. The first has to do with the number and type of stress resultants associated with any cut:

Space beams. When a space beam is cut an internal member force vector and an internal moment vector are both generated (six scalar quantities). *Plane beams.* When a plane beam is cut, an internal member force vector and a scalar moment are both generated (three scalar quantities).

Both of these statements can be argued from the equations of statics for a cantilever beam. For example, since the cut space beam must be in force equilibrium, the possibility of arbitrary loads requires the existence of an arbitrary internal force *vector* associated with the cut. Similarly, moment equilibrium requires the existence of an internal moment vector associated with the cut. For plane beams this argument degenerates to the existence of two scalar force components and one scalar moment component at any cut.

There is one final point to be made here which gives rise to the concept of a *local coordinate system*. It is not enough to simply compute the internal forces and moments under discussion, they must be computed in a proper coordinate system if they are to be useful when member stresses are to be computed. For example, it is common to divide the axial force by the member area to obtain the 'axial stress'. In this case, strength of materials requires that this axial force be the component of the internal force vector which is parallel to the member centerline. Just any component will not do!

Fig. 1.12 attempts to make this point graphically. The straight beam first of all tends to blur the issues since it is so simple, but the curved beam makes the point clearly. With respect to statics alone it would be adequate to simply compute the global components F_x, F_y, and M but from the structures point of view that is simply not adequate. To be useful it is necessary to compute the components T, V, and M of the 'local coordinate system'. The figure attempts to generalize this discussion to the case of space beams.

The idea of a rotation matrix comes about naturally through the necessity of computing stress resultants in special coordinate systems. That is the point of Example 1.4. Given the reactions, it is a relatively simple matter to compute a set of stress resultants at the upper point (say the cut).

Example 1.4   Given the circular beam segment and the reactions, find T, V, and M.

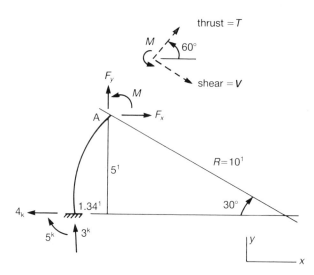

If the forces are to sum to zero,

$$F_x = 4^k, \qquad F_y = -3^k$$

If the moments are to sum to zero about point A, then

$$4 \times 5 + 3 \times 1.34 + 5 - M = 0 \Rightarrow M = 29.02 \text{ k}'$$

Using the rotation matrix,

(*continued on page 31*)

Plane beam (straight)

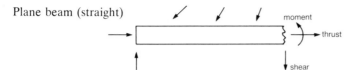

(local coordinate system = global coordinate system)

Plane beam (curved)

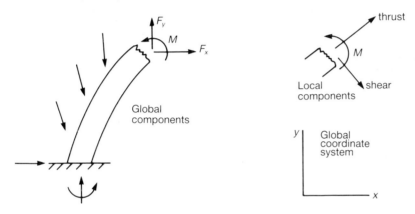

Three-dimensional beams

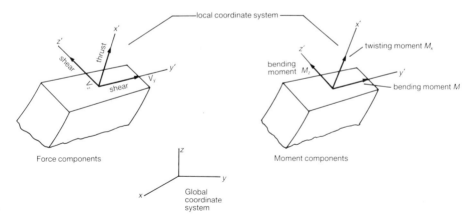

Note – In the local coordinate system:

(1) The x' axis is parallel to the beam centerline

(2) The y' and z' axes are aligned with the principal axes of the beam cross-section.

Fig. 1.12 – Internal member forces (stress resultants) in 'local' and 'global' coordinate systems.

$$\begin{bmatrix} A'_x \\ A'_y \end{bmatrix} = \begin{bmatrix} \cos\theta & \sin\theta \\ -\sin\theta & \cos\theta \end{bmatrix} \begin{bmatrix} A_x \\ A_y \end{bmatrix} \Rightarrow \begin{bmatrix} \text{thrust} \\ -\text{shear} \end{bmatrix} = \begin{bmatrix} \cos 60° & \sin 60° \\ -\sin 60° & \cos 60° \end{bmatrix} \begin{bmatrix} 4 \\ -3 \end{bmatrix}$$

$$= \begin{bmatrix} -0.6^k \\ -4.96^k \end{bmatrix}$$

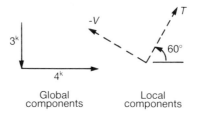

Global   Local
components   components

Alternative: Compute T and V using projections

$T = 2 - 2.6 = -0.6^k$

$V = +3.46 \times 1.5 = 4.96^k$

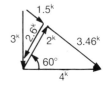

Force equilibrium requires horizontal and vertical force components of 4^k and 3^k respectively. The moment at A can be computed by writing the moment equilibrium equations about point A as indicated in the figure.

The point is that it is necessary to make one additional step and compute the shear and thrust. Since the shear and thrust are statically equivalent to the components in the glocal coordinate system, both sets of forces must simply be different components of the same force vector at A. It follows that the rotation matrix (see Appendix 1) can be used to compute these force components as indicated in this example. The alternative of computing the shear and the thrust using projections is also indicated.

Example 1.5 shows the difficulties which arise when working in three dimensions. Given, for example, the reactions at point A, the equations of statics can be used to compute to forces and moments required at B for equilibrium. It is then again necessary to compute the internal stress resultants at this point in the local coordinate system. That is done using the rotation matrix.

Three-dimensional problems are sufficiently complex to make a more organized approach using a rotation matrix a necessity. Attempting to use scalar projections (which is of course identical to using a rotation matrix) tends to be confusing in three dimensions largely due to difficulties in visualizing three-dimensional quantities.

Example 1.5   Helical beam (stair girder slab).

Given the forces at end A,

$$F_Z = (0, 10^k, 0)$$
$$\mathbf{M_A} = (10^{k'}, 0, 0)$$

Find the stress resultants at point B.

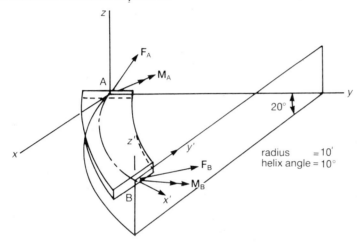

Step 1. Find the reactions at point B in the global coordinate system.

Sum forces: $\mathbf{F_A} + \mathbf{F_B} = 0$ $\mathbf{F_B} = -\mathbf{F_A} = (0, -10, 0)$

Sum moments about point B:

$$\mathbf{M_A} + \mathbf{M_B} + \mathbf{r} \times \mathbf{F_A} = 0 \Rightarrow \mathbf{M_B} = -\mathbf{M_A} - \mathbf{r} \times \mathbf{F_A}$$

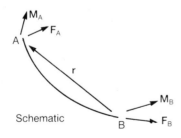

Schematic

Components of **r**:

$$r_x = -10 \cdot \sin 20° = -3.41'$$
$$r_y = -10 \cdot (1 - \cos 20°) = -0.61' \quad \Rightarrow \quad \mathbf{M_B} = (-16.15, 0, 34.1)$$
$$r_z = -10 \cdot \frac{20}{180} \pi \cdot \tan 10° = -0.615'$$

Step 2. Find the rotation matrix (see Appendix 1)

Method 1. Compound rotations
Rotate the global coordinate system into the local coordinate system at point B. First rotate $20°$ about the z-axis then rotate $-10°$ about the y-axis. The composite rotation matrix R is then

$$R = \begin{bmatrix} \cos(-10°) & 0 & -\sin(-10°) \\ 0 & 1 & 0 \\ \sin(-10°) & 0 & \cos(-10°) \end{bmatrix} \begin{bmatrix} \cos 20° & \sin 20° & \\ -\sin 20° & \cos 20° & 0 \\ 0 & 0 & 1 \end{bmatrix}$$

$$= \begin{bmatrix} 0.925 & 0.336 & 0.174 \\ -0.341 & 0.937 & 0 \\ -0.163 & -0.059 & 0.985 \end{bmatrix}$$

Method 2. Base vectors

The unit vectors of the local coordinate system form the rows of the rotation matrix.

$$\mathbf{j}' = (-\sin 20°, \cos 20°, 0) = (-0.342, 0.939, 0)$$

$\mathbf{i}'$ has the *direction* $(-\cos 20°, \sin 20°, \tan 10°)$

normalize $\mathbf{i}'$ for $\mathbf{i}' = (0.926, 0.337, 0.174)$

$$\mathbf{k}' = \mathbf{i}' \times \mathbf{j}' = (-0.163, -0.059, 0.985) \qquad \text{check!}$$

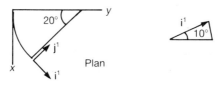

Plan

Step 3. Multiply $\mathbf{M_B}$ and $\mathbf{F_B}$ by the rotation matrix.

$$\begin{bmatrix} \text{thrust} \\ \text{shear}_y \\ \text{shear}_z \end{bmatrix} = F'_B = RF_B = \begin{bmatrix} 0.925 & 0.336 & 0.174 \\ -0.341 & 0.937 & 0 \\ -0.163 & -0.059 & 0.985 \end{bmatrix} \begin{bmatrix} 0 \\ -10 \\ 0 \end{bmatrix}$$

$$= \begin{bmatrix} -3.35^k \\ -9.39^k \\ 0.592^k \end{bmatrix}$$

$$\begin{bmatrix} \text{twisting moment} \\ \text{bending moment}_y \\ \text{bending moment}_z \end{bmatrix} = M'_B = RM_B = \begin{bmatrix} -9 \ k' \\ 5.18 \ k' \\ 36.1 \ k' \end{bmatrix}$$

1.5 EXERCISES

1 Show that $\mathbf{A} \times \mathbf{B} = -\mathbf{B} \times \mathbf{A}$.

2 Show that $\mathbf{A} \times \mathbf{B} \perp \mathbf{A}$, $\mathbf{B}$ or that $\mathbf{A} \cdot (\mathbf{A} \times \mathbf{B}) = \mathbf{B} \cdot (\mathbf{A} \times \mathbf{B}) = 0$.

3 Show that $\mathbf{A} \times (\mathbf{B} \times \mathbf{C}) = \mathbf{B}(\mathbf{A} \cdot \mathbf{C}) - \mathbf{C}(\mathbf{A} \cdot \mathbf{B})$.

4 Show that $\mathbf{A} \cdot (\mathbf{B} \times \mathbf{C}) = (\mathbf{A} \times \mathbf{B}) \cdot \mathbf{C} = -\mathbf{A} \cdot (\mathbf{C} \times \mathbf{B})$.

5 Show that $(\mathbf{A} \times \mathbf{B}) \times \mathbf{C} = \mathbf{B}(\mathbf{C} \cdot \mathbf{A}) - \mathbf{A}(\mathbf{C} \cdot \mathbf{B})$.

6 Show that $\mathbf{A} \times [\mathbf{A} \times (\mathbf{A} \times \mathbf{B})] = (\mathbf{A} \cdot \mathbf{A})(\mathbf{B} \times \mathbf{A})$.

7 Show that the scalar product of two vectors is equal to the cosine of the angle between the two vectors multiplied by the product of their absolute values.

8 Given a force which is allowed to move along a given line. Show that the moment of this force about a given point is independent of this motion.

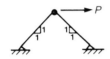

9 Find the reactions for the plane truss shown.

10
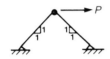

10 Find the reactions for the space truss shown.

11
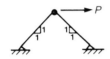

11 Find the reactions for the circular arch shown. (Find them in terms of shear and thrust.)

12 For the helix of Example 1.5 under its own weight, assume point A to be free and again compute the stress resultants at B. (Use a $4' \times 8''$ concrete slab.)

Solution Exercise 12.

Step 1. Compute the weight of the beam segment.

Let ρ be the weight per foot of slab. α = helix angle = $10°$.

$$\text{Weight of } 20° \text{ segment} = \int_l \rho \, ds = \int_0^{20°} \rho r \, d\theta \sec \alpha$$

$$= \rho r \sec \alpha \, 20\pi/180.$$

Concrete weighs 145 lb/ft³ ⇒

$8'' \times 4'$ slab weighs $\frac{8}{12} \times 4 \times 145 = 387$ lb/ft

Total slab weight $= 387 \times 10' \times 1.015 \times 20 \times \pi/180$

$= \underline{1371}$ lb $= W$

Step 2. Compute the centroid of the helical segment.

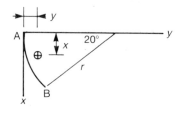

Equations of a helix:

$$x = r \sin \theta \qquad y = r(1 - \cos \theta) \qquad z = r\theta \tan \alpha$$

$$W \times \bar{x} = \int_0^{20°} x \rho \, ds = \int_0^{20°} \rho r \, d\theta \sec \alpha \cdot r \cdot \sin \theta$$

$$= \rho r^2 \sec \alpha \int_0^{20°} \sin \theta \, d\theta = \rho r^2 \sec \alpha (1 - \cos 20°)$$

$$= 387 \times 10^2 \times 1.015 \times 0.063 = 2369 \text{ ft lb}$$

$$\bar{x} = \frac{W \times \bar{x}}{W} = \frac{2369}{1371} = \underline{1.728'}$$

$$W \times \bar{y} = \int_0^{20°} \rho \, dsy = \int_0^{20°} \rho r \, d\theta \sec \alpha \, r(1 - \cos \theta)$$

$$= \rho r^2 \sec \alpha (\theta - \sin \theta) \, \big|_0^{20°}$$

$$= 387 \times 10^2 \times 1.015 \times \left(\frac{20\pi}{180} - \sin 20° \right) = 276.7$$

$$\bar{y} = \frac{276.7}{1371} = \underline{0.2019'}$$

Step 3. Compute the reactions of B in the global coordinate system.

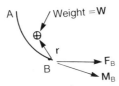

Forces sum to zero: $\mathbf{W} + \mathbf{F}_B = 0$

$\mathbf{F}_B = -\mathbf{W} = \underline{(0, 0, +1371\ )}$

Moments sum to zero about point B
$\mathbf{M}_B + \mathbf{r} \times \mathbf{W} = 0 \qquad \mathbf{M}_B = -\mathbf{r} \times \mathbf{W}$

$\mathbf{r} = (-(r \sin \theta - \bar{x})_1 - (r(1 - \cos \theta) - \bar{y}, 0)$
$= (-1.692, -0.4012, 0)$

$\mathbf{M}_B = -\mathbf{r} \times \mathbf{w} = (0.4012 \times 1371, 1.692 \times 1371, 0)$
$= \underline{(-550, 2370, 0)}\ \text{ft lb}$

Step 4. Compute stress resultants in the local coordinate system. (See Example 1.5 for unit vectors)

thrust $= \mathbf{i}' \cdot \mathbf{F}_B = 238$ lb
$\text{shear}_y = \mathbf{j}' \cdot \mathbf{F}_B = 0$
$\text{shear}_z = \mathbf{k}' \cdot \mathbf{F}_B = 1350$ lb

twisting moment $= \mathbf{i}' \cdot \mathbf{M}_B = 272$ lb$'$
$\text{bending moment}_y = \mathbf{j}' \cdot \mathbf{M}_B = 2368$ lb$'$
$\text{bending moment}_z = \mathbf{k}' \cdot \mathbf{M}_B = -48$ lb$'$

13 The cantilever beam shown has a parabolic shape with vertical tangent at point A rather than the circular shape of Example 1.4. Find T, V, M.

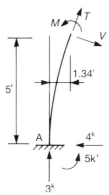

14 Compute the reactions F_1 and F_2 for the plane truss shown. (See Example 1.3.)

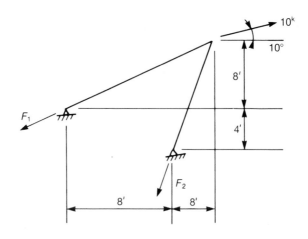

15 Compute the reactions F_1, F_2, F_3 for the space truss shown. (See Example 1.2.)

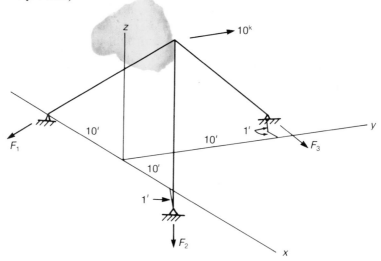

2. Statically Determinate Structures

2.1 INTRODUCTION

This relatively long chapter is concerned with a topic which is basic to any introductory text on structures: the analysis of statically determinate structures. A statically determinate structure is, first of all, a structure for which the internal member forces can be computed *uniquely* using only the given external loads and the equations of statics. These computed internal member forces are then *independent* of the materials of which the structure has been constructed, the effects of temperature and settlement, . . .

In order to develop a perspective for statically determinate structures, a parallel with the equations of linear algebra is useful. First note that the equations of statics as they are used in this book, no matter how they are constructed, are linear equations in the unknown forces and moments. (This is due to the fact that these equations are always written in terms of the given initial configuration rather than the deformed configuration which would be more correct.) Now, given a set of linear equations $Ax = b$, there are three alternatives with regard to solution type as indicated in Fig. 2.1. For each of these alternatives there is a structural classification:

(1) *Statically determinate structures*. When the equations of statics imply a unique set of internal member forces and moments for an arbitrary set of external loads, a structure is said to be statically determinate. The word 'arbitrary' is important here. A structure may perform well under one set of loads but still be technically unstable. For example, the unstable truss of Fig. 2.1 performs well under an upward vertical load but cannot carry a horizontal load component and for that reason is regarded to be unstable.

(2) *Statically indeterminate structures*. The case of the statically indeterminate structure of Fig. 2.1 is equally interesting and is in fact central to this book. It was implied in Chapter 1 (Example 1.3) that the two-

bar truss is generally statically determinate. When a bar is added as was done in Fig. 2.1, the resulting structure is statically indeterminate and the equations of statics no longer provide a unique solution for the member forces. In fact, the force in the bar which has been added may be selected arbitrarily and the equations of statics still satisfied since the added bar, with respect to the remaining two bars, has the effect of an external load.

Linear algebra	*Structural analysis*
A system of linear equations may have:	The equations of statics for a given structure are:
(1) A unique solution.	(1) Statically determinate.
(2) Many solutions.	(2) Statically indeterminate.
(3) No solutions.	(3) Geometrically unstable.

Examples:

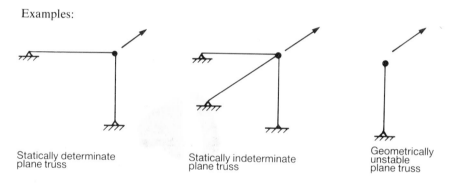

Statically determinate
plane truss

Statically indeterminate
plane truss

Geometrically
unstable
plane truss

Fig. 2.1 – Classification of structures.

(3) *Geometrically unstable structures.* The case of an unstable structure is somewhat pathological and will not be dealt with in much detail in this book. One example has already been cited in Fig. 2.1 and another will be discussed later in this chapter. Unstable structures are as a class somewhat counter-productive since they are neither designed nor built.

One additional comment on the non-uniqueness of the equilibrium equations is in order. In general, it can be expected that within linear theory a given set of external loads will produce a unique set of internal member forces. The fact that the equilibrium equations can possess non-unique solutions simply implies that the structure is not completely described by them. The remedy, which will be discussed at length in Chapter 4, will be to introduce member stiffness properties into the description of the structure. The additional requirement that the deformed structure must fit together will then supply sufficient additional conditions to guarantee the unique solution which might have been argued on physical grounds.

By way of summary then, statically determinate structures are introduced at this point for several reasons:

(1) *Simplicity*. Statically determinate structures form a simple subclass of structures.

(2) *A step toward the general case*. In Chapter 4 it will be shown how statically indeterminate structures can be reduced to statically determinate structures which then require certain corrections which depend upon member stiffness properties.

(3) *Construction*. For many practical reasons, statically determinate structures are easy to build and have for that reason alone become quite common.

This chapter moves slowly with increasing complexity through the analysis of statically determinate trusses, frames, and some simple cases of shells. A subsequent chapter will present methods for the computation of displacements in statically determinate structures which then form a basis for the analysis of statically indeterminate structures.

2.2 STATICALLY DETERMINATE TRUSSES

Technically, a truss is a structure made up of pin-connected members which is loaded only at its joints. The result of this definition is that with regard to statics, each member can be described by a single scalar quantity known as the 'bar force'. That is the point of Fig. 2.2.

Since the bar in Fig. 2.2 is pin-ended, no moments are shown at its ends. The bar is then acted upon by two force vectors. The equations of

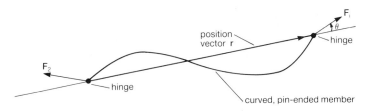

Fig. 2.2 – Pin-ended member.

equilibrium will now be used to show that (a) these forces are equal and opposite and (b) they must be parallel to a line passing through the pinned ends of the member. Since the forces must sum to zero for any rigid body in equilibrium, it follows that

$$\mathbf{F}_1 + \mathbf{F}_2 = 0 \qquad \text{(the reactions are equal and opposite)} \qquad (2.1)$$

Writing moment equilibrium about the left end of the bar requires that

$$\mathbf{r} \times \mathbf{F}_1 = 0 \qquad \text{(this implies that } \mathbf{r} \parallel \mathbf{F}_1) \qquad (2.2)$$

Since $|\mathbf{r} \times F_1| = |\mathbf{r}| \cdot |\mathbf{F}_1| \cdot \sin \theta$, it follows that in the non-trivial case, $\sin \theta = 0 \Rightarrow \theta = 0$ or that the force $\mathbf{F}_1$ must lie along a line between the two hinges. This gives rise to the concept of a bar force. (See Fig. 2.3.)

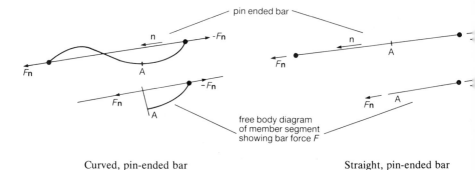

Curved, pin-ended bar   Straight, pin-ended bar

Fig. 2.3 – Concept of a bar force.

Since the force on the end of a pin-ended bar must lie along the line between the pins, on any free body diagram of a bar which has been cut, the internal force must also lie along this line and can therefore be represented by a scalar times a unit vector. This scalar is called the bar force and is taken to be positive when the member is in tension.

For trusses there is a simple, necessary test for statical determinacy. The node equilibrium equations first provide a set of independent equations; there are $3j$ of these for space trusses and $2j$ for plane trusses, when j is the number of joints. The number of unknowns in either case is equal to the sum of the number of reactions r and the number of bar forces b. For a truss to be statically determinate, the number of equations must equal the number of unknowns

	Space truss	*Plane truss*
Number of equations	$3j$	$2j$
Number of unknowns	$b + r$	$b + r$
Statical determinancy	$3j = b + r$	$2j = b + r$

Fig. 2.4 shows how these relationships apply to two specific examples. What this type of analysis does not show is whether or not a structure is

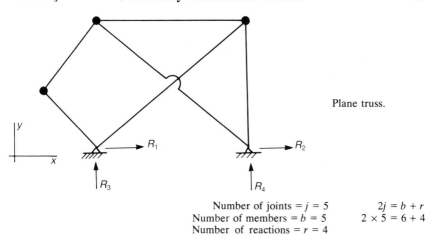

Plane truss.

Number of joints = j = 5 $2j = b + r$
Number of members = b = 5 $2 \times 5 = 6 + 4$
Number of reactions = r = 4

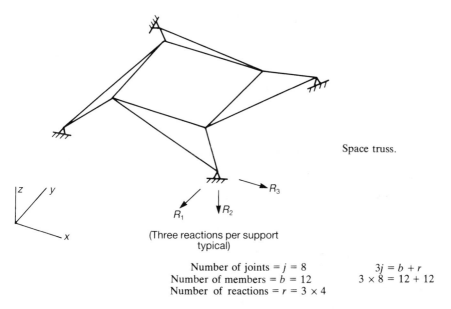

Space truss.

(Three reactions per support typical)

Number of joints = j = 8 $3j = b + r$
Number of members = b = 12 $3 \times 8 = 12 + 12$
Number of reactions = $r = 3 \times 4$

Fig. 2.4 – Counting equations and unknowns for statically determinate trusses.

geometrically unstable. In fact, an example below shows a case in which these relationships are satisfied, but the structure is unstable in a rather unpredictable manner. But this is not common.

One final comment concerning real trusses. While it is uncommon today to find a structure which actually contains hinges, trusses remain a common model for many classes of structures. Fig. 2.5 is a case in point.

Fig. 2.5 – Chesapeake and Delaware Canal bridge (lift bridge) Kirkwood–Mt Pleasant, Delaware. (Courtesy of Gerard Fox, Howard, Needles, Tammen, and Bergendoff, NY City.)

This bridge has heavy riveted joints which are probably closer to fixed joints than they are to hinges, but the structure itself has been analyzed and designed as if its members were pin-connected. The fact that structures like this can be modeled as trusses has a lot to do with member stiffness and less to do with member connections. If the members themselves are sufficiently flexible (think of a string) then it does not matter how they are connected: *the moments which develop will be small.* In other cases such as wooden domes it is the connections themselves which are flexible while the members themselves may be quite stiff. In either case members are not to be subjected to significant bending if a truss model is to be valid.

Finally there are practical reasons for not using actual hinges in structures such as the fact that (a) they are expensive to make and (b) they do not work well (unless they are carefully maintained).

2.2.1 Plane Trusses

Fig. 2.6 shows some of the common configurations of plane trusses which have developed over the years with specific applications and specific

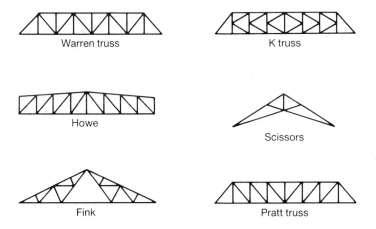

Fig. 2.6 – Some common types of plane trusses.

materials in mind. In some respects, the engineer is a prisoner of his design requirements. For example, a parallel chord truss under gravity load acts much like a beam with the upper chord in compression (like the upper flange of a beam) and the lower in tension and little can be done about it. On the other hand, by changing the direction of the diagonals they can be forced to act either like tension members (common in steel roof trusses) or like compression members. Plane trusses can in fact be found in many possible configuration depending on the functional requirements of building shape and bridge span and upon material considerations. The discussion is endless and not especially relevant to this text.

This section is concerned with developing methods for the analysis of plane trusses which can be performed by hand and which add to the understanding of how trusses function. It is noted in passing that the most systematic way to approach a truss is to write the joint equilibrium equations as simultaneous linear equations in the unknown bar forces and reactions. These equations can then be solved at low cost using a digital computer. While elegant and systematic, this is not a fruitful way to proceed when working by hand since humans do not perform this type of task well. The approach taken in this section will be to explain some basic principles of truss behavior and then show how these principles can be used to solve trusses without solving large systems of equations.

(1) *Trusses with three reactions.* Since there are three equilibrium equations for a plane rigid body, when a plane truss has three reactions it is always possible to compute these directly. As a practical matter, if these reactions are to be computed it is generally a good idea to compute them first.

Example 2.1 indicates a typical structure in which it is required to compute a three reaction components R_1, R_2, R_3. If a moment

---

Example 2.1   A plane truss with three unknown reactions.

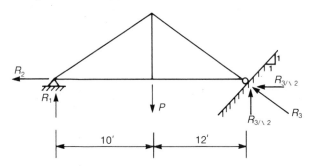

Moments must sum to zero about the left support

$$10P - 22R_3/\sqrt{2} = 0 \quad \Rightarrow \quad \underline{R_3 = \sqrt{2}\,\tfrac{10}{22}P}$$

Forces must sum to zero in the horizontal and vertical directions

$$R_2 + \frac{R_3}{\sqrt{2}} = 0 \quad \Rightarrow \quad \underline{R_2 = -\tfrac{10}{22}P}$$

$$R_1 - P + R_3/\sqrt{2} = 0 \quad \Rightarrow \quad \underline{R_1 = \tfrac{12}{22}P}$$

---

equation is written about the left support, R_3 can be computed directly; then horizontal and vertical equilibrium can be used to compute R_1 and R_2.

(2) *The two-bar truss.* It should by now be clear that the plane two-bar truss is statically determinate unless the bars are collinear. (It is then geometrically unstable.) This implies that:

(a) Any two-bar joint in a plane truss may be analyzed directly and that

(b) The two bar forces of an unloaded two-bar joint must be zero. (This is true since the equilibrium equations become homogeneous. That is, if the system $Ax = b$ possess a unique inverse A^{-1} then $b = 0 \Rightarrow x = 0$.)

Both of these points are illustrated in Example 2.2. As indicated, joint B can be solved directly as a two-bar truss for bar forces F_a, F_b; since joint c is unloaded it can be argued without further analysis that $F_c = F_d = 0$.

Example 2.2   Find bar forces F_a, F_b, F_c, F_d

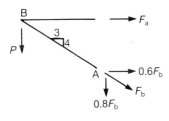

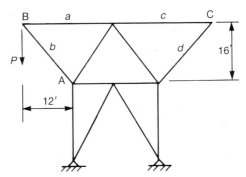

Free body diagram of joint B

Vertical forces sum to zero: $P + 0.8\, F_b = 0 \Rightarrow \underline{F_b = -\tfrac{5}{4}P}$

Horizontal forces sum to zero: $F_a + 0.6\, F_b = 0 \Rightarrow \underline{F_a = \tfrac{3}{4}P}$

$\underline{F_c = F_d = 0}$(unloaded two-bar joint)

---

(3) *Collinear bars at a three-bar joint.* If two bars at a three-bar joint are collinear, the force in the remaining bar may be computed directly. If the joint is unloaded the remaining bar force must be zero. (See Fig. 2.7 and Example 2.3.)

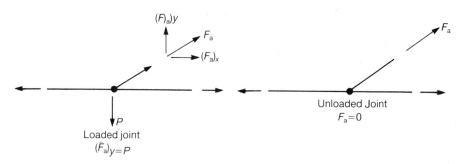

Fig. 2.7 – Collinear bars.

Example 2.3   Find F_a, F_b

$$(F_a)_y = F_a \sin 60° = 10 \Rightarrow \underline{F_a = 11.54 \text{ lb}}$$

$\underline{F_b = 0}$ (unloaded joint)

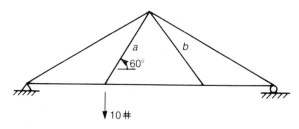

---

(4) *Bar force components.* If any component of a bar force is known, the bar force itself may be computed directly. (See Fig. 2.8 and Example 2.3.)

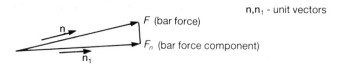

Fig. 2.8 – Bar force component. By definition: $F_n = n_1 \cdot (\mathbf{Fn}) \Rightarrow F = F_n/(\mathbf{n}_1 \cdot \mathbf{n})$.

Example 2.4 discusses a final two-bar truss. Note the two approaches used: either the straightforward application of force equilibrium or a combined approach using both moment and force equilibrium may be used.

Example 2.4 Find F_1, F_2

Force equations

$\mathbf{n}_1 = (-1/\sqrt{2}, 1/\sqrt{2})$ $F_1\, 1/\sqrt{2} + F_2 \times 0.8 = 2$

$\mathbf{n}_2 = (-0.8, -0.6)$ $F_1\, 1/\sqrt{2} - F_2 \times 0.6 = 7$ $\Rightarrow$ $1.4\, F_2 = -5$ $\Rightarrow F_2 = -3.58$

$$\frac{F_1}{\sqrt{2}} = 2 - 0.8F_2 \Rightarrow F_1 = 6.85$$

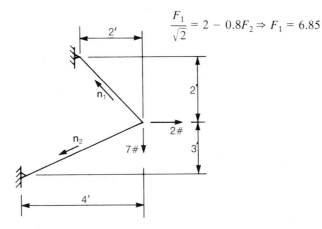

Alternative method of solution:

Moment equation (about lower support)

$$\frac{F_1}{\sqrt{2}} \times 5 + \frac{F_1}{\sqrt{2}} \times 2 = 7 \times 4 + 2 \times 3 = 34 \Rightarrow F_1 = 6.85$$

Summing vertical forces

$$0.6 \times F_2 = -7 + \frac{6.85}{\sqrt{2}} = -2.15 \Rightarrow F_2 = -3.58$$

2.2.1.1 The Method of Joints

The method of joints makes systematic use of the equations of force equilibrium in moving from joint to joint throughout a truss computing bar forces. The idea is that (like the two-bar truss) any truss joint which contains only two unknown bar forces can be solved for these bar forces using the equations of force equilibrium. As a practical matter note again that:

(a) When possible, reactions should be computed first.
(b) When any component of a bar force is known, the bar force itself can be computed.

Example 2.5 shows how the method of joints can be applied to a truss of modest size.

Example 2.5 Use of the method of joints.

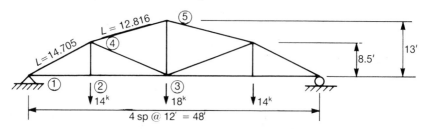

Step 1. Compute member lengths (required for projections)
Step 2. Symmetry: Reactions $= 14 + 9 = 23^k$
Only need to solve half the truss

Joint 1.

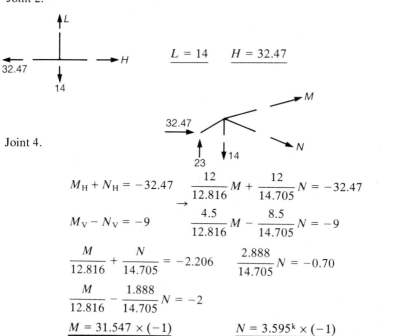

$G_V = -23 \Rightarrow$

$G = -23 \times \dfrac{14.205}{8.5}$

$= -39.79$

$G_H = G_V \times \dfrac{12}{8.5} = -32.47 \Rightarrow F = 32.47$

Joint 2.

$L = 14 \qquad H = 32.47$

Joint 4.

$M_H + N_H = -32.47 \qquad \dfrac{12}{12.816} M + \dfrac{12}{14.705} N = -32.47$

$M_V - N_V = -9 \qquad \dfrac{4.5}{12.816} M - \dfrac{8.5}{14.705} N = -9$

$\dfrac{M}{12.816} + \dfrac{N}{14.705} = -2.206 \qquad \dfrac{2.888}{14.705} N = -0.70$

$\dfrac{M}{12.816} - \dfrac{1.888}{14.705} N = -2$

$M = 31.547 \times (-1) \qquad N = 3.595^k \times (-1)$

Joint 5.

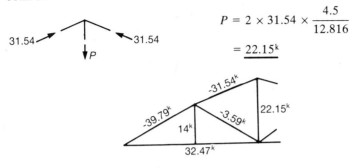

$$P = 2 \times 31.54 \times \frac{4.5}{12.816}$$
$$= \underline{22.15^k}$$

Bar force summary

From a practical standpoint, it is sometimes convenient to make a relatively large, scale drawing of the truss to be analyzed and write force components directly upon it as if the joints themselves were free body diagrams. That approach is indicated in Example 2.6 and is highly recommended in general.

Example 2.6   Method of joints. Find the bar forces.

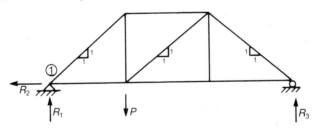

Step 1. Compute reactions

Sum moments about the left support

$$3R_3 - 1 . P = 0 \Rightarrow R_3 = P/3$$

Sum forces $R_1 = \frac{2}{3}P$ $R_2 = 0$

Step 2. Solve joint free body diagrams

Typical joint free body diagram

Joint 1

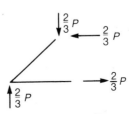

Bar forces shown on sketch of truss:

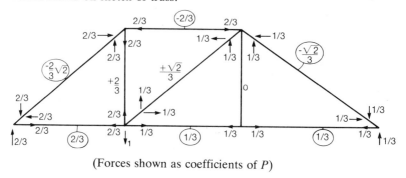

(Forces shown as coefficients of P)

2.2.1.2 The Method of Sections

When you wish to find all the bar forces in a truss, the method of joints provides a systematic way of doing so. There are other cases in which it may only be necessary to find some individual bar forces. When this can be done through the clever use of free body diagrams it is called the *method of sections*. Example 2.7 indicates some uses of this method. In this case moment equations are written about the intersection points of some bar forces allowing the other bar forces to be computed directly.

Example 2.7   Use of the method of sections
(This truss has already been discussed in Example 2.5.)

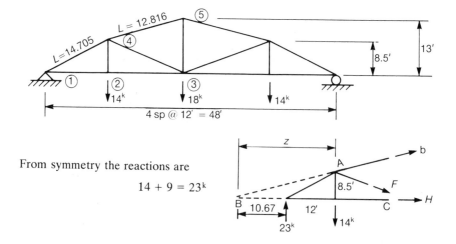

From symmetry the reactions are

$$14 + 9 = 23^k$$

Moments about point A will give the bar force H

directly: $8.5 \times H = 23 \times 12$ or $\underline{H = 32.47^k}$

Moments about point B will give the bar force F directly:

Compute the location of part B $\dfrac{8.5}{z} = \dfrac{4.5}{12}$ $z = 22.67'$

Moments about B: $F_V \times (22.67 + 12) + 14 \times 22.67 - 13 \times 10.67 = 0$

$$F_V = -2.07 \;\Rightarrow\; F = -2.07 \times \frac{14.705}{8.5} = \underline{-3.59^k}$$

Moments about part C will give the bar force G directly:

$$G_H \times 13 + 23 \times 24 - 14 \times 12 = 0 \qquad G_H = -29.54^k$$

$$G = -29.54 \times \frac{12.82}{12} = \underline{-31.55^k}$$

Another application of the method of sections occurs in parallel chord trusses where a vertical section cutting the diagonals can be used with vertical force equilibrium to compute directly the forces in the diagonals (Fig. 2.9). Returning to Example 2.6, the vertical force component in the center diagonal must therefore equal the right reaction which is 1/3. The method of sections is another one of these cases in which an experienced engineer can be extremely effective by knowing the proper use of the proper free body diagram. It is at the same time impossible to do justice to the method with a few simple examples. The reader will in any case see the method used repeatedly throughout this text.

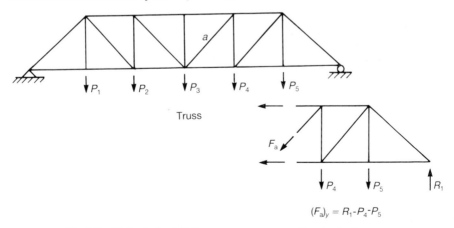

Truss

Fig. 2.9 – Method of sections.   Free body diagram

$(F_a)_y = R_1 - P_4 - P_5$

2.2.1.3 Geometric Instability

This section contains the final example in this text of a structure which is geometrically unstable. As suggested earlier, in the case of a geometrically unstable structure the equilibrium equations have no solution. (That is obvious in the case of Fig. 2.10. Since the bar forces have no vertical

Truss   Free body diagram

Fig. 2.10 – Geometrically unstable truss.

components it is impossible to satisfy vertical equilibrium.) On the other hand, it is more common to think of an unstable structure kinematically in terms of collapse. That is, an unstable structure possesses the possibility of a rigid body motion which implies potential collapse. While not quite in the spirit of the material of this text, there is a beautiful theorem of linear algebra (see Appendix 2) which describes this force/displacement duality of geometric instability. Translated into structural terms it states that

Either the equilibrium equations have a solution for arbitrary load or there exists a collapse mechanism.

The dangerous part of geometric instability is that it is not always obviously predictable. That is the point of this section. It will first of all be shown that the Wichert truss is generally stable and can be analyzed without any particular difficulty. It will then be shown that in a special, largely unpredictable case the Wichert truss is geometrically unstable.

Fig. 2.11 attempts to show the general configuration of the Wichert

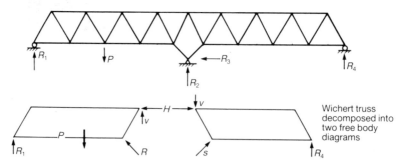

Wichert truss decomposed into two free body diagrams

Fig. 2.11 – The Wichert truss.

truss which uses a rather peculiar looking central panel with standard trusses on either side. While the Wichert truss is statically determinate, because of its configuration it behaves like an indeterminate structure to the extent that the load P on one span produces a reaction R_4 on the other span. In any case the analysis of the Wichert truss is interesting. First, note that the truss of Fig. 2.11 satisfies the necessary condition for statical determinancy since there are 36 bars, four reactions, and 20 joints and therefore

$$2j = b + r \quad \text{or} \quad 2 \times 20 = 36 + 4$$

Still there is no obvious starting point for the analysis since there are four reactions and no two-bar joints to be solved. A common procedure in situations like this is to begin to take the structure apart in order to use internal structural conditions. (This will be done again later for the case of the three-hinged arch.) That is done in Fig. 2.11 and it appears that the decomposed structure may be manageable. It contains two rigid bodies which supply $2 \times 3 = 6$ equilibrium equations and there are six unknowns R_1, R_4, H, V, R, S.

Two specific cases will be presented here in detail. First, Example 2.8

Example 2.8 An unstable Wichert truss.

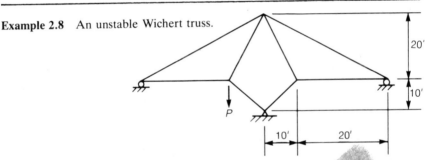

$R/\sqrt{2} \times 30 + R_1 \times 30 - 10P = 0$ (moments about H, V on the left).

$S/\sqrt{2} \times 30 + R_4 \times 30 \qquad\quad = 0$ (moments about H, V on the right).

$R_1 - P + R/\sqrt{2} + S/\sqrt{2} + R_4 = 0$ (vertical forces sum to zero).

$R/\sqrt{2} - S/\sqrt{2} \qquad\qquad\quad = 0$ (horizontal forces sum to zero).

In matrix form

$$
\begin{bmatrix}
1/\sqrt{2} & 0 & 1 & 0 \\
0 & 1/\sqrt{2} & 0 & 1 \\
1/\sqrt{2} & 1/\sqrt{2} & 1 & 1 \\
1 & -1 & 0 & 0
\end{bmatrix}
\begin{bmatrix}
R \\
S \\
R_1 \\
R_4
\end{bmatrix}
=
\begin{bmatrix}
\dfrac{10P}{30} \\
0 \\
P \\
0
\end{bmatrix}
$$

describes a Wichert truss which will now be shown to be unstable. Rather than using the six equations mentioned above, this example uses four equations in the four unknowns R, S, R_1, R_4 which are obtained by simply avoiding the terms H and V. That can be done by writing the moment equations about their point of application and by putting the two pieces together before writing the equations of horizontal and vertical equilibrium. The fact that the structure is unstable can best be seen from the matrix form of the equilibrium equations in this example. Since row three in the matrix is a linear combination of rows 1 and 2 these equations are clearly singular. In kinematic terms, the particular dimensions of this problem allow a kind of 'squashing' rigid body motion in the structure which is not obvious.

A more typical Wichert truss is described in Example 2.9. The analysis begins with the computation of the four unknowns R, S, R_1, R_4 as described in Example 2.9 except that horizontal equilibrium is immediately used to set $R = S$. Having these reactions, the method of joints is used to compute the bar forces.

Example 2.9   Analysis of a Wichert truss.
Step 1. Find the reactions R, S, R_1, R_4 (see Fig. 2.11).

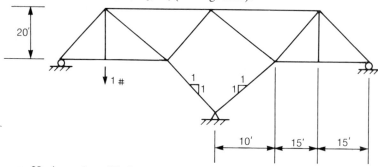

Horizontal equilibrium $\Rightarrow R = S$

Vertical equilibrium $\quad \Rightarrow R_1 + \dfrac{2R}{\sqrt{2}} + R = 1$

Moment equilibrium for the left piece $\Rightarrow \quad 40R_1 + \dfrac{R}{\sqrt{2}} \cdot 30 = 25$

Moment equilibrium for the right piece $\Rightarrow \quad 40R_4 + \dfrac{R}{\sqrt{2}} \cdot 30 = 0$

Solving: $\quad R \;=\; \dfrac{15}{20} \cdot \sqrt{2} = 1.0607$

$$R_1 = \frac{1}{16} = 0.0625$$

$$R_4 = -\frac{9}{16} = -0.5625$$

Step 2. Compute the bar forces using the method of joints.

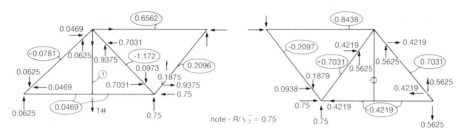

2.2.2 Space Trusses

For all intents and purposes, this section parallels Section 2.2.1 which is concerned with plane trusses. The reader is invited to compare these two sections.

(1) *Trusses with six reactions.* Since there are six equations of equilibrium for a rigid body in space, when a three-dimensional structure has six unknown reaction components it should be possible to compute these components directly unless the structure is unstable. An example of a structure with six reaction components is discussed in Example 2.12 but it is not a situation commonly encountered.

(2) *The three-bar truss.* It should by now be clear that the three-bar truss is statically determinate unless the three bars (possibly when extended) all pass through the same line. Then, unless the structure is geometrically unstable:

   (a) Any three-bar space truss can be analyzed directly.
   (b) The three-bar forces of an unloaded three-bar joint must be zero.

   Both of these points are illustrated in Example 2.10. The forces at joint A may be immediately declared to be zero since A is an unloaded three-bar joint. Since joint B is a three-bar joint it may be solved directly.

(3) *One bar out of the plane.* Fig. 2.12 shows the case of a joint at which a number of bars lie in one plane and one bar lies out of the plane. Writing equilibrium normal to this plane allows the out of plane bar force F to be computed as

$$\mathbf{n}_p \cdot \mathbf{P} + \mathbf{n}_p \cdot (\mathbf{n}F) = 0$$

or

$$F = -\frac{\mathbf{n}_p \cdot \mathbf{P}}{\mathbf{n}_p \cdot \mathbf{n}} \qquad (2.3)$$

When the applied force **P** has no component normal to the plane or

Example 2.10 Space truss.

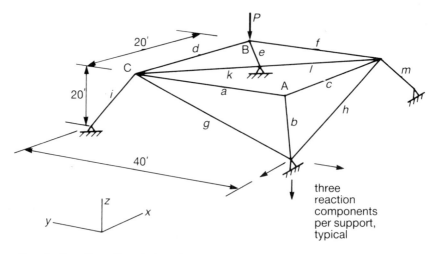

Step 1. $F_a = F_b = F_c = 0$ since joint A is an unloaded three-bar joint.

Step 2. Since bars d and f lie in the $x - y$ plane,

$$P + (F_e)_z = 0 \Rightarrow (F_e)_z = -P \Rightarrow F_e = -P\sqrt{6}/2$$

(the slope of the corner bars is 1, 1, 2)

Step 3. Because of this slope $(F_3)_x = (F_e)_y = -P/2$. x and y equilibrium of joint B then implies that

$$F_d = F_f = -\frac{P}{2}$$

Step 4. $F_g = 0$ since bar g is out of the plane of bars i, k, d. Similarly $F_{h = 0.}$

Step 5. Sum moments about the axis shown in the free body diagram.

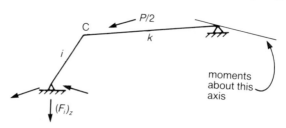

$$(F_i)_z \cdot 40 + \frac{P}{2} \cdot 20 = 0$$

$$(F_i)_z = -\frac{P}{4} \Rightarrow F_i = -\frac{P}{4} \frac{\sqrt{6}}{2}$$

using the slope again. Similarly

$$F_m = -\frac{P}{4} \frac{\sqrt{6}}{2}$$

Step 6. Equilibrium at joint 6 in the z direction $\Rightarrow (F_i)_z = (F_k)_z$ or $+P/4 = (F_k)_z$, but the slope of bar k is 30, 10, 20

$$F_k = +\frac{P}{4} \frac{\sqrt{14}}{2} \quad \text{similarly} \quad F_l = +\frac{P}{4} \frac{\sqrt{14}}{2}$$

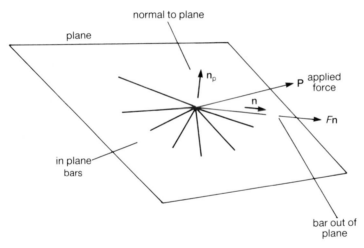

Forces in a direction normal to the plane must sum to zero

$$\mathbf{n_p} \cdot \mathbf{P} + \mathbf{n_p} \cdot (\mathbf{n}F) = 0$$

$$F = \frac{\mathbf{n_p} \cdot \mathbf{P}}{\mathbf{n_p} \cdot \mathbf{n}}$$

Fig. 2.12   Bars in a plane.

when the joint is unloaded, this bar force must be zero. Again, see Example 2.10 for some specific applications of this property.

(4) *Bar force components.* In three was well as two dimensions, when a component of a bar force is known the force itself can be computed directly.

Example 2.11 discusses the final three-bar space truss of this section. Note again that two approaches are possible. Either the force equations of equilibrium may be written for the loaded joint and solved as simultaneous equations or moment equations may be used three times to construct three

---

Example 2.11   Find the bar forces F_1, F_2, F_3.

Step 1. Compute unit vectors
Step 2. Write equilibrium equations
Step 3. Solve for bar forces

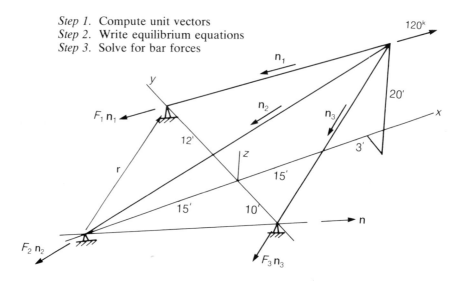

Bar	Projections			Length	Unit vector comps		
1	−15	15	−20	29.1	−0.517	0.517	−0.688
2	−30	3	−20	36.1	−0.831	0.083	−0.554
3	−15	−7	−20	25.9	−0.579	−0.271	−0.772
n	15	−10	0	18.1	0.829	−0.552	0
r	15	12	0	19.2			

Joint equilibrium

$$F_1\mathbf{n}_1 + F_2\mathbf{n}_2 + F_3\mathbf{n}_3 + \mathbf{P} = 0$$

$$
\begin{aligned}
-0.517\,F_1 & \quad -0.831\,F_2 & \quad -0.579\,F_3 + 120 = 0 \\
0.517\,F_1 & \quad +0.083\,F_2 & \quad -0.271\,F_3 \quad\quad\;\; = 0 \\
-0.688\,F_1 & \quad -0.554\,F_2 & \quad -0.772\,F_3 \quad\quad\;\; = 0
\end{aligned}
$$

Eliminate F_1: Eq. (2.1) and Eq. (2.2) $-0.748\,F_2 - 0.850\,F_3 + 120 = 0$

$$\text{Eq. 2 + Eq. 3} \times \left(-\frac{0.517}{0.688} \right) \quad -0.415\,F_2 - 0.002\,F_3 + 120 = 0$$

$-0.333\,F_2 - 0.852\,F_3 = 0$
or $F_2 = -2.56\,F_3$.
   Finally

$$F_3 = -113^k \qquad F_2 = 290^k \qquad F_1 = -108^k$$

Alternative method: Find F_1 by taking moments about line n

$$F_1 = -\frac{\mathbf{n} \cdot (\bar{\mathbf{r}} \times \mathbf{P})}{\mathbf{n} \cdot (\mathbf{r} \times \mathbf{n}_1)} = \frac{-(0.552 \times 0.554 \times 36.1 \times 120)}{0.829[12 \times (-0.688)] - (-0.552) \times 15 \times (-0.688)}$$

$$= \frac{1330}{-0.688 \times 18.31} = -108^k \qquad \text{check}$$

where
$$\mathbf{P} = (120, 0, 0)$$
$$\bar{\mathbf{r}} = -L_2\mathbf{n}_2 = -36.1(-0.831, 0.083, -0.554)$$

---

uncoupled equations. This latter technique is illustrated in Fig. 2.13. If, for example, a moment equation is written about a line through two of the supports, the third bar force can be computed directly since the first two bar forces clearly have no moment about this line. Then

$$\mathbf{n} \cdot \mathbf{r} \times (\mathbf{n}_1 F_1) + \mathbf{n} \cdot (\mathbf{F} \times \mathbf{P}) = 0$$

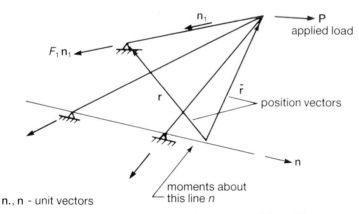

Since line n possess through two of the reactions, the moments of F_1 and $\mathbf{P}$ must sum to zero about line n

$$\mathbf{n} \cdot \mathbf{r} \times (\mathbf{n}_1 F_1) + \mathbf{n} \cdot (\bar{\mathbf{r}} \times \mathbf{P}) = 0$$

or

$$F_1 = -\frac{\mathbf{n} \cdot (\bar{\mathbf{r}} \times \mathbf{P})}{\mathbf{n} \cdot (\mathbf{r} \times \mathbf{n}_1)}$$

Fig. 2.13   Computing a bar force using a moment equation.

or

$$F_1 = -\frac{\mathbf{n} \cdot (\bar{\mathbf{r}} \times \mathbf{P})}{\mathbf{n} \cdot (\mathbf{r} \times \mathbf{n}_1)} \tag{2.4}$$

This process must be repeated two more times to compute the other two bar forces.

2.2.2.1 The Method of Joints

As in the case of plane trusses, it is sometimes possible to move from joint to joint through a three-dimensional truss finding joints at which there are only three unknown bar forces or reactions, computing these unknowns using the equations of joint equilibrium, and then moving on to another joint until all the unknowns have been computed. Example 2.12 indicates how this can be done in a particular case.

Example 2.12   Assume from symmetry that $R_6 = R_7$. Find all the bar forces.

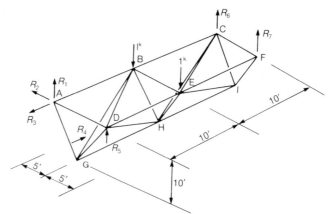

Solution by the method of joints:

(1) By symmetry all reactions must be $\frac{1}{2}$ lb.

(2) At joint A, bar AG must carry the $\frac{1}{2}$ lb reaction.

   therefore $F_{AG} = \frac{1}{2} \backslash \frac{\overline{5}}{2} = \underline{0.559 \text{ lb}}$. Since bars AG, GB, GH all lie in a plane $F_{DG} = 0$. $F_{AD} = -\frac{1}{4}$ lb.

(3) At joint D, bar DH must carry the reaction of $\frac{1}{2}$ or $F_{HD} = 0.5 \times \frac{3}{2} = \underline{0.75 \text{ lb}}$.

(4) At joint G, F_{GH} must balance the $\frac{1}{2}$ lb vertical reaction or $F_{GH} = \frac{1}{2}$ lb.

(5) Equilibrium of joint D ⇒ $\underline{F_{DB} = 0}$.

(6) The other results follow directly.

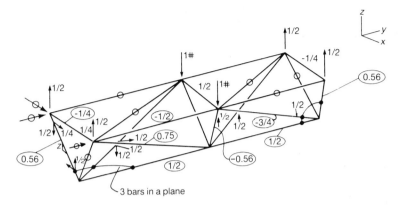

2.2.2.2 The Method of Sections
As in the case of plane trusses, the method of sections for space trusses involves the clever use of free body diagrams. How this can be done has already been discussed in several examples above. No additional cases will be represented here.

2.2.2.3 The Schwedler Dome
Example 2.10 describes a simple three-dimensional truss which can be constructed by starting with two tripods and adding tripods in a rather obvious manner. The Schwedler dome is something of a generalization of this exercise. (See Fig. 2.14.) The analysis of the Schwedler dome is relatively straightforward. At any joint, say joint 'a' of this figure, three bars lie in a plane with one bar out of the plane. As described above in Eq. (2.3), it is possible to compute the remaining bar force by summing

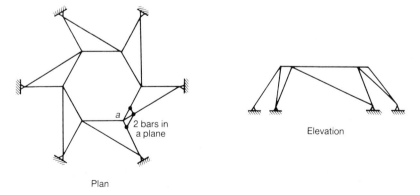

Fig. 2.14 – Single layer Schwedler dome.

Example 2.13   Analysis of a Schwedler dome.

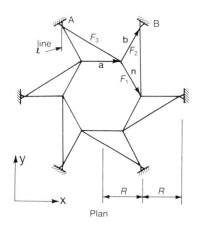

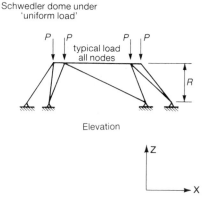

Schwedler dome under
'uniform load'

Elevation

Plan

Normal to the plane defined by **a** and **b**

$$\mathbf{a} = (1, 0, 0) \qquad \mathbf{b} \cong (R \cos 60°, R \sin 60°, -R)$$

$$= (0.5, 0.866, -1) \quad \text{length} = \sqrt{2}$$

$$\mathbf{a} \times \mathbf{b} = (0, 1, 0.866); \text{ normalize to } \mathbf{n} = (0, 0.7559, 0.655)$$

Solve for F_1:

$$F_1 = -\frac{\mathbf{P} \cdot \mathbf{n}}{\mathbf{n}_1 \cdot \mathbf{n}} \qquad \mathbf{P} = (0, 0, -P)$$

$$\mathbf{n} \cdot \mathbf{P} = -0.655P \qquad \mathbf{n}_1 \sim (R/2, -0.866R, 0)$$
$$= (0.5, -0.866, 0)$$

$$\mathbf{n}_1 \cdot \mathbf{n} = -0.6546 \qquad F_1 = -\frac{-0.655P}{-0.655} = \underline{-P}$$

Solve for F_2 by taking moments about line l

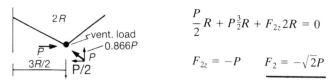

vent. load
0.866P

$$\frac{P}{2}R + P\tfrac{3}{2}R + F_{2z}2R = 0$$

$$F_{2z} = -P \qquad F_2 = -\sqrt{2}P$$

Since F_2 balances the applied load $F_3 = 0$. Solution checks symmetry!

forces normal to the plane. This process is illustrated in Example 2.13. Finally note that layers of this type may be stacked up to form more complex domes.

2.3 FRAMES

Within the terminology used in this text, the general class of skeletal structures is the frame (sometimes 'rigid frame'). It is formed by joining together beams in an arbitrary manner and can be made unstable by, for example, inserting hinges in it. Having said that, the instability of frames will not be discussed further. The truss then is a very special case of the frame. Or put another way, a skeletal structure which is not a truss is a frame.

By the analysis of a frame is meant determining the stress resultants at every point in the structure. In the truss it was adequate to describe each member by giving its bar force, a single scalar quantity. In plane frames, for example, the moments, shears, and thrusts vary from point to point and are commonly described by plots or *stress resultant diagrams.* Since a plane frame has three stress resultants at each point it is then necessary to construct three stress resultant diagrams for each member; similarly, a space frame requires six stress resultant diagrams for each member.

Two approaches are common when computing stress resultants. It is most elegant to construct the appropriate differential equations governing the stress resultants of interest and then solve them for particular cases of loads and boundary conditions. As a practical matter, these differential equations turn out to be cumbersome and most practical analysis is done by 'cutting' structures at points of interest and using the equations of equilibrium for a rigid body to compute the stress resultants. In either case the results must of course be identical.

Two final comments are appropriate concerning the differential equations which are derived below. The derivations presented are, first of all, heuristic and the reader should be aware that they can be described more carefully using Taylor series expansions. This brings up the second point. If it is assumed in the derivations that the functions involved are continuous, then how is it possible to handle practical cases of point loads and moments? The common way of dealing with point loads is to break the solution on either side of the load and then match up these solutions with appropriate discontinuity conditions. Another approach is described in Appendix 3 which shows how discontinuous functions can be used in these cases. But the use of discontinuous functions is really of more theoretical interest than practical interest. For practical problem solving, discontinuous functions add little to what can be done without them. They do, however, have a formal utility in that they allow differential equations and expressions for work and energy to be written simply without having to bother with a lot of special cases. The argument being that the special cases of point loads and moments can be subsumed within the discussion of distributed loads as limiting cases.

2.3.1 Equilibrium Equations of Straight Plane Beams

The straight plane beam is the most simple example of a frame and one of the most common structural elements. Given a straight beam acted upon by lateral load $w(x)$ and a distributed axial load $t(x)$, the differential equations of equilibrium which relate these loads to the moment M, shear V, and thrust T, are derived in Fig. 2.15 and simply listed here for convenience as

$$\left.\begin{array}{l} \dfrac{dV}{dx} = -w \\[2mm] \dfrac{dM}{dx} = V \end{array}\right\} \Rightarrow \dfrac{d^2M}{dx^2} = -w$$

$$\dfrac{dT}{dx} = -t$$

(2.5)

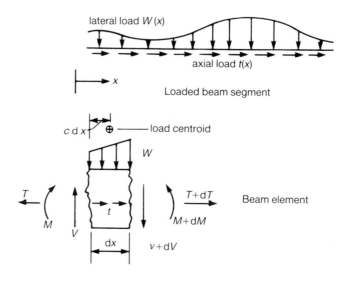

lateral load $W(x)$

axial load $t(x)$

Loaded beam segment

$c\,dx$ —— load centroid

w

T $T+dT$ Beam element

M V t $M+dM$

dx $v+dV$

(1) Sum vertical forces:

$$w\,dx + dV = 0 \Rightarrow dV/dx = -w$$

(2) Sum moments:

$$w\,dx\,.\,c\,dx + (V + dV)\,dx - dM = 0$$
$$\Rightarrow dM/dx = V$$

(3) Sum horizontal forces:

$$t\,dx + dT = 0 \Rightarrow dT/dx = -t$$

Fig. 2.15 – Equilibrium of a plane straight beam.

Note that it is most common in simple applications to find the thrust T to be zero; in any case the thrust is uncoupled from the moment and shear in Eq. (2.5) and can be handled separately if need be.

Example 2.14 discusses three particular cases of beam analysis:

(1) *Uniform load*. When w is constant the solution is quite simple to obtain. The equilibrium equations require two integrations and thus need two boundary conditions. Shear and moment diagrams are plotted. In this example it is also shown how the structure can be solved by 'cutting' it and applying the equations of equilibrium to the rigid bodies which result.

(2) *Sinusoidal load*. This is a somewhat unusual case in which the differential equations are probably easier to deal with than the cut structure depending of course on the integrals which are available.

(3) *Concentrated load*. Without recourse to the discontinuous functions of Appendix 3, this beam must be solved in two parts and the parts matched at the load P. On each side of the load P the lateral load w is zero which implies that the shear is constant and that the moment is linear. A free body diagram must then be used to derive 'discontinuity conditions' which must be satisfied when the final solution is assembled.

Example 2.14   Find the moment and shear.

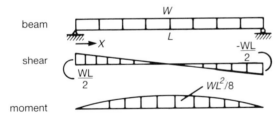

Case 1. Uniform load

$$dV/dx = -W \Rightarrow V = -Wx + c_1 = dM/dx$$

$$M = -Wx^2/2 + c_1 x + c_2$$

Boundary conditions: $m = 0 @ x = 0 \Rightarrow c_2 = 0$
$\qquad\qquad\qquad\quad M = 0 @ x = L \Rightarrow c_1 = Wl/2$

Finally, $V = W(l/2 - x)$

$$M = \frac{Wx}{z}(l - x)$$

Alternative method. Compute reactions and then 'cut' the structure. Reactions both $Wl/2$ by symmetry.

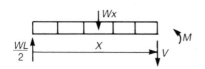

Sum force: $V = \dfrac{Wl}{2} - Wx$

Sum moments: $M = \dfrac{Wl}{2}x - Wx\,x/2$

Case 2. Sinusoidal load

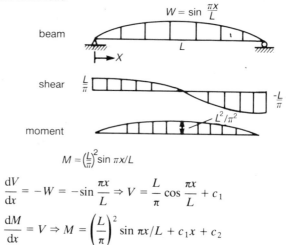

$$M = \left(\tfrac{L}{\pi}\right)^2 \sin \pi x/L$$

$$\frac{dV}{dx} = -W = -\sin \frac{\pi x}{L} \Rightarrow V = \frac{L}{\pi} \cos \frac{\pi x}{L} + c_1$$

$$\frac{dM}{dx} = V \Rightarrow M = \left(\frac{L}{\pi}\right)^2 \sin \pi x/L + c_1 x + c_2$$

Boundary conditions: $m = 0$ @ $x = 0, L \Rightarrow c_1 = c_2 = 0$

Finally, $V = \dfrac{L}{\pi} \cos \pi x/L$

Case 3. Point load

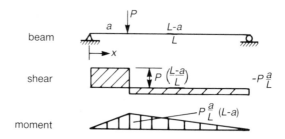

On either side of the load P, $w = 0 \Rightarrow$ shear is constant and the moment linear. Match solutions at the load P. Equilibrium requires that $M^+ = M^-$ (moment continuous) and that $(V^+ - V^{45f}) + P = 0$.

Boundary conditions: $m = 0$ @ $x = 0, L$.

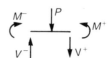

Finally:

$$0 \leqslant x \leqslant a \qquad\qquad a \leqslant x \leqslant L$$

$$V = P\left(\frac{L-a}{L}\right) \qquad V = =P\frac{a}{L}$$

$$M = P\left(\frac{L-a}{L}\right)x \qquad M = P\frac{a}{L}(L-x)$$

2.3.2 Plane Frames and Arches

The work-horse method for solving frames and arches is simply to cut members at points of interest and then use the equations of equilibrium to compute the internal stress resultants. When a structure is statically determinate this approach must yield a solution once the reactions have been computed. Although a more elegant approach, solving the differential equations of equilibrium is not a common procedure in practical situations. With the arbitrary load and arbitrary geometry commonly encountered in design, it is usually more productive to compute stress resultants directly rather than doing so through the differential equations. In any case the differential equations for an arbitrarily curved beam will be returned to below for the case of three-dimensional beams.

Example 2.15 illustrates the typical steps taken during a frame analysis. First, the reactions are computed. In this case since there are three unknown reactions and three equations of equilibrium for a rigid body, these reactions can be computed directly. In some cases such as the three-hinged arch discussed below there can be four reactions to compute. In any case, given the reactions, enough free body diagrams must be taken (cuts to the structure must be made) to describe the stress resultants at every point. In this particular example, four free body diagrams are required.

Example 2.15 Compute and plot stress resultant diagrams.

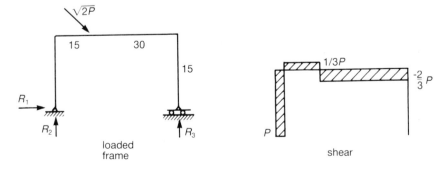

loaded frame

shear

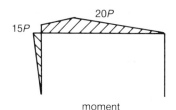

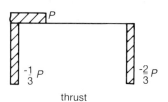

moment   thrust

(1) Use moment equation about the left support to compute the right equation.

$$15P + 15P - R_3 \times 45 = 0 \qquad R_3 = \tfrac{2}{3}.$$

(2) Sum forces for other reactions

$$R_1 = -P \qquad R_2 = P/3$$

(3) Use free body diagrams to compute stress resultants.

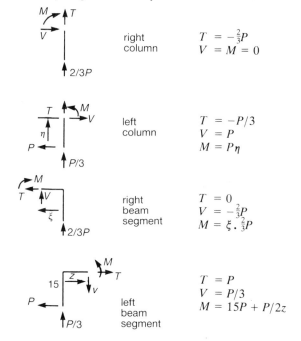

right column $\qquad$ $T = -\tfrac{2}{3}P$
$V = M = 0$

left column $\qquad$ $T = -P/3$
$V = P$
$M = P\eta$

right beam segment $\qquad$ $T = 0$
$V = -\tfrac{2}{3}P$
$M = \xi \cdot \tfrac{2}{3}P$

left beam segment $\qquad$ $T = P$
$V = P/3$
$M = 15P + P/2z$

Example 2.16 is included here to pick up the pace a little. While it is always important to be methodical it is necessary at some point to begin to anticipate matters somewhat or the details of performing an analysis simply become overwhelming. Toward this end note that:

Example 2.16 Compute and plot stress resultant diagrams.

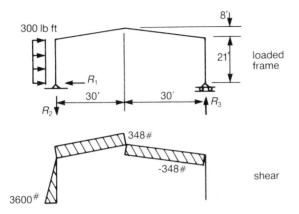

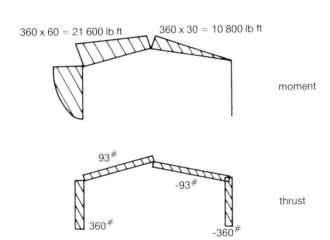

(1) Compute reactions

Moments about left support $\Rightarrow R_3 = 360$ lb
Force equilibrium $\Rightarrow R_1 = 3600$ lb
$$R_2 = R_3 = 360 \text{ lb}$$

(2) Compute stress resultants

Right column: $T = -360$ lb $V = M = 0$

Right beam segment: $T = -93$ lb
                                 $V = -348$

$\tan^{-1}(8/30) = 14.93°$ $M = 360 . \xi$
$360 . \sin 14.93 = 93$ lb
$360 . \cos 14.93 = 348$ lb

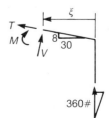

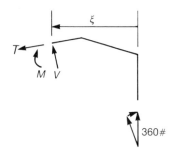

Left beam sigment:

$T = 93$ lb
$V = -348$ lb
$M = 360 . \xi$

Left column segment:

$T = 360$ lb
$V = 3600 - 300x$
$M = 3600x - 300x^2/2$

(a) For a straight segment subjected only to concentrated loads, the moment diagram must be piecewise linear and the shear and thrust diagrams piecewise constant. In these cases the concern lies primarily with the endpoints of the segments. (Note that a 'kink' in a frame is something like a loaded point, since shear and thrust are typically discontinuous there.)

(b) Much of the detail of the stress resultant diagrams can be anticipated directly. For example, the moment along the right column must be zero, the moment at the peak of the frame is $360 \times 30 = 10\ 800$, the moment at the left knee is $360 \times 60 = 21\ 600$, and the moment along the left column is parabolic (since the loading is uniform).

(c) In terms of understanding the results, it is a good idea when possible to plot the stress resultant diagrams on a sketch of the structure. That procedure is followed in this text.

(d) Unless point couples are applied, the moment diagram must be continuous. This is even true at frame corners and can be useful information when plotting and checking moment diagrams.

2.3.2.1 The Three-hinged Arch

There is no strong distinction between a frame and an arch. Their different connotations are as much historical as anything else and have something to do with the type of structure under discussion. For example, a 'rigid frame' used in a building might be called an 'arch' when used in a bridge. Generally in an arch bridge you are attempting to turn lateral load into arch thrust avoiding bending as much as possible; in a rigid frame roof you tend to accept the loads and simply design for the resulting bending moment. There will be more of this discussion later in the chapter on cable analysis. The interest in this section lies with curved structural members.

Fig. 2.16 shows a three-hinged arch schematically; this arch is appropriate to this chapter since it is a statically determinate structure.

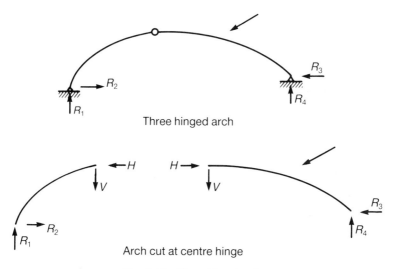

Three hinged arch

Arch cut at centre hinge

Fig. 2.16 – Three-hinged arch.

That can be argued in two ways. First, a single frame bent (Fig. 2.17) without hinges is statically indeterminate to the third degree since it becomes statically determinate when fully cut and the full cut requires that three stress resultants be set to zero. The other argument follows the discussion of the Wichert truss (Fig. 2.11). Since the three-hinged arch has four reactions, these cannot be computed directly from the three equations of equilibrium for a rigid body. However, when the structure is cut at the hinge into two rigid bodies, a case of six equations (two rigid bodies) in the six unknowns R_1, R_2, R_3, R_4, H, V results.

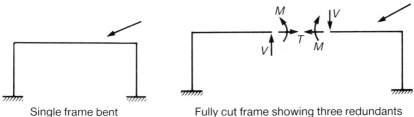

Single frame bent   Fully cut frame showing three redundants

Fig. 2.17 – Rigid frame.

Example 2.17 shows a symmetric three-hinged semicircular arch acted upon by a single load. A free body diagram of the left half structure is first used to obtain a relationship between H_1 and R_1. Once this has been established the reactions can be computed. Finally, two free body diagrams are required to describe the stress resultants fully.

Example 2.17   Compute and plot the stress resultants for the three-hinged semi-circular arch shown.

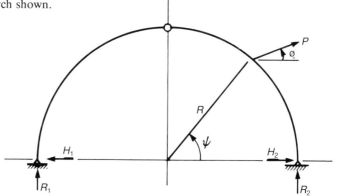

Step 1.   Relate R_1 to H_1. Sum moments about center hinge.

$$R_1 . R + H_1 . R = 0 \qquad H_1 = -R_1$$

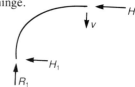

Step 2.   Compute reactions. Sum moments about left support

$$(P . \cos \phi)(R \sin \psi) - (P \sin' \phi) R(1 + \cos \psi) - R_2 . 2R = 0$$

$$R_2 = \frac{P}{2} (\cos \phi . \sin \psi - \sin \phi(1 + \cos \psi)$$

Sum vertical force components

$$R_1 + R_2 + P \sin \phi = 0$$

Sum horizontal force components

$$-H_1 + H_2 + P \cos \phi = 0$$

Step 3.   Compute stress resultants.

Left free body diagram.

$$F_2 = -R_1$$
$$F_1 = H_1$$

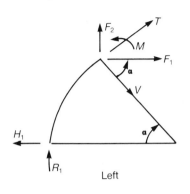

Left

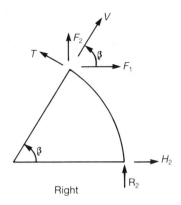

Right

Rotate for T and V.

$$V = F_1 \cos(-\alpha) + F_2 \sin(-\alpha)$$
$$T = -F_1 \sin(-\alpha) + F_2 \cos(-\alpha)$$
$$M = R[H_1 \sin \alpha + R_1(1 - \cos \alpha)]$$

Similar results for right free body diagram.

Special case. (symmetric) $\phi = \psi = 90°$

$$0 \leqslant \alpha \leqslant 90°$$

$$V = \frac{P}{2}(\cos \alpha - \sin \alpha)$$

$$T = \frac{P}{2}(\sin \alpha + \cos \alpha)$$

$$M = \frac{PR}{2}(\sin \alpha + \cos \alpha - 1)$$

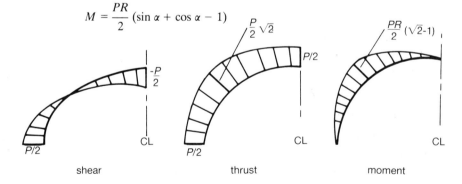

shear   thrust   moment

Example 2.18 describes the analysis of a three-hinged parabolic arch under uniform load. In this classic case both the shear and moment are zero throughout the arch and the lateral load is carried completely by thrust. While this is an ideal situation from the point of view of design (since structures are most efficient when carrying axial load), it has limited application. First, most structures cannot simply be designed for a single loading condition and most loading conditions are not uniform. Furthermore, while the analysis presented in Example 2.18 is valid for any value of the rise h, for small values of h it would be expected that deflections eventually become more important with decreasing h and that a nonlinear theory should then be used in preference to the present linear analysis.

In Chapter 6 it will be shown that a cable under uniform load takes the shape of a parabola. This parabolic arch is them simply an inverted cable, at least as far as statics is concerned.

Example 2.18 Three-hinged parabolic arch under uniform load.

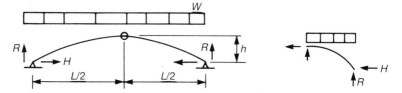

Moments about left support:

$$RL = wlL/2 = 0 \qquad R = wL/2$$

Moments about the hinge

$$Hh - \frac{RL}{2} + \frac{wl}{2}\frac{L}{4} = 0 \qquad H = \frac{wL^2}{8h}$$

Equation of a parabola:

$$y = ax^2 + bx + c \qquad \begin{aligned} y = 0 \ @\ x = 0 &\Rightarrow c = 0 \\ y = 0 \ @\ x = L &\Rightarrow 0 = aL^2 + bL + c \Rightarrow b = -aL \\ y = h \ @\ x = L/2 &\Rightarrow h = a\left(\frac{L}{2}\right)^2 + bh/2 + c \end{aligned}$$

finally

$$y = 4h\left\{\frac{x}{L} - \left(\frac{x}{L}\right)^2\right\} \qquad y' = -\frac{4h}{L}\left[2\left(\frac{x}{L}\right) - 1\right]$$

Free body diagram for internal forces:

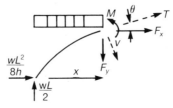

$$F_x = -\frac{wL^2}{8h} \quad F_y = \frac{wL}{2} - wx$$

$$M = \frac{wL}{2}x - \frac{wL^2}{8h}y - \frac{wx^2}{2}$$

$$= \frac{wL}{2}x - \frac{wL^2}{8h}\left[4h\left\{\frac{x}{L} - \left(\frac{x}{L}\right)^2\right\}\right] - \frac{wx^2}{2} = O$$

Transform for T and V

$$\left(\theta = \tan^{-1}\left(-\frac{4h}{L}\left(\frac{2x}{L} - 1\right)\right)\right)$$

$$T = F_x\cos\theta - F_y\sin\theta = -\frac{wL^2}{8h}\cos\theta - \left[\frac{wL}{2} - wx\right]\sin\theta$$

$$Y = F_y\cos\theta + F_x\sin\theta = \left[\frac{wL}{2} - wx\right]\cos\theta - \frac{wL^2}{8h}\sin\theta$$

$$V = \frac{\left(\frac{wL}{2} - wx\right)L^2 + \frac{wL^2}{8L}4h(2x - L)}{\sqrt{16h^2(4x^2 - 4 \times L + L^2) + L^4}} = O$$

$$T = \frac{-w\frac{L^2}{8h}L^2 + \left[\frac{wL}{2} - wx\right]4h(2x - L)}{\sqrt{16h^2(4x^2 - 4 \times L + L^2) + L^4}}$$

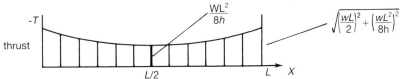

2.3.3 **Space Frames**
In terms of practical analysis, the space frame is a direct extension of the material of the preceding section. That is, once reactions have been computed, internal stress resultants can be computed by cutting the structure at the points of interest. Only in this case it is necessary to compute and plot six stress resultants at every point.

Example 2.19 describes a straight cantilever beam in space loaded at its end. This is a rather trivial example since the local coordinate system is in this case the same as the global coordinate system and the analysis reduces to what are essentially plane problems.

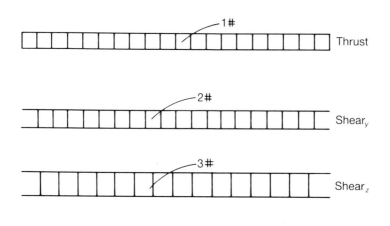

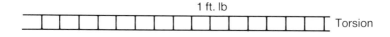

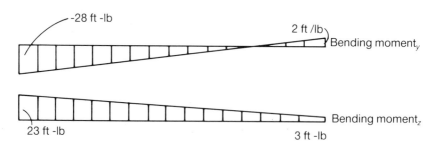

Example 2.19 A cantilever beam in space.

Example 2.20 describes a cantilever stair girder under its own weight. Note that in three dimensions the coordinates of the centerline of a beam do not fully describe the beam since its cross-section can be twisted about its centerline. However when a stair girder is specified, the upper surface of the beam must be horizontal in a direction normal to the beam centerline. This fact then completes the description of the beam and thus the description of the local coordinate system. Because the member is kinked at point B, two free body diagrams are required. Otherwise the analysis is straightforward.

Example 2.20 Stair girder ABC is cantilevered from point C where it is fixed. Its weight is 500 lb/ft. Compute and plot the stress resultants.

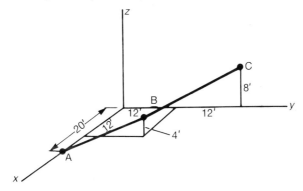

Step 1. Find the rotation matrices

Bar AB

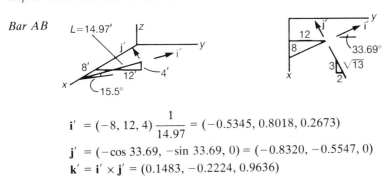

$$\mathbf{i'} = (-8, 12, 4)\frac{1}{14.97} = (-0.5345, 0.8018, 0.2673)$$

$$\mathbf{j'} = (-\cos 33.69, -\sin 33.69, 0) = (-0.8320, -0.5547, 0)$$

$$\mathbf{k'} = \mathbf{i'} \times \mathbf{j'} = (0.1483, -0.2224, 0.9636)$$

Check using rotations about coordinate axes.

Rotate (1) $90° + 33.69°$ about z-axis then (2) $-15.5°$ about y-axis

$\theta = -15.5°$ $\theta = 123.69°$

$$R = \begin{bmatrix} \cos\theta & 0 & -\sin\theta \\ 0 & 1 & 0 \\ \sin\theta & 0 & \cos\theta \end{bmatrix} \cdot \begin{bmatrix} \cos\theta & \sin\theta & 0 \\ -\sin\theta & \cos\theta & 0 \\ 0 & 0 & 1 \end{bmatrix} = \begin{bmatrix} -0.5345 & 0.8018 & 0.2672 \\ -0.8320 & -0.5547 & 0 \\ 0.1482 & -0.2223 & 0.9636 \end{bmatrix}$$

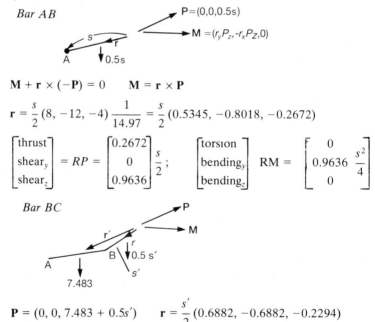

Bar BC

$$\mathbf{i'} = (-12, 12, 4)\frac{1}{17.44} = (-0.6882, 0.6882, 0.2294)$$

$$\mathbf{j'} = (-\cos 45°, -\sin 45°, 0) = (-0.7071, -0.7071, 0)$$

$$\mathbf{k'} = (0.1622, -0.1622, 0.9732)$$

Check using rotations about coordinate axes.

Rotate (1) 135° about the z-axis then (2) $-13.26°$ about the y-axis

$\theta = -13.26°$ $\theta = 135°$

$$R = \begin{bmatrix} \cos\theta & 0 & -\sin\theta \\ 0 & 1 & 0 \\ \sin\theta & 0 & \cos\theta \end{bmatrix} \cdot \begin{bmatrix} \cos\theta & \sin\theta & 0 \\ -\sin\theta & \cos\theta & 0 \\ 0 & 0 & 1 \end{bmatrix} = \begin{bmatrix} -0.6882 & 0.6882 & 0.2294 \\ -0.7071 & -0.7071 & 0 \\ 0.1622 & -0.1622 & 0.9732 \end{bmatrix}$$

Step 2. Compute stress resultants

Bar AB $P = (0, 0, 0.5s)$

$M = (r_y P_z, -r_x P_z, 0)$

$$\mathbf{M} + \mathbf{r} \times (-\mathbf{P}) = 0 \qquad \mathbf{M} = \mathbf{r} \times \mathbf{P}$$

$$\mathbf{r} = \frac{s}{2}(8, -12, -4)\frac{1}{14.97} = \frac{s}{2}(0.5345, -0.8018, -0.2672)$$

$$\begin{bmatrix} \text{thrust} \\ \text{shear}_y \\ \text{shear}_z \end{bmatrix} = RP = \begin{bmatrix} 0.2672 \\ 0 \\ 0.9636 \end{bmatrix}\frac{s}{2} \; ; \qquad \begin{bmatrix} \text{torsion} \\ \text{bending}_y \\ \text{bending}_z \end{bmatrix} \quad RM = \begin{bmatrix} 0 \\ 0.9636 \\ 0 \end{bmatrix}\frac{s^2}{4}$$

Bar BC

$$\mathbf{P} = (0, 0, 7.483 + 0.5s') \qquad \mathbf{r} = \frac{s'}{2}(0.6882, -0.6882, -0.2294)$$

$$\mathbf{r'} = \mathbf{r} + (4, -6, -2)$$

$$\mathbf{M} + \mathbf{r} \times (0, 0, -0.5s') + \mathbf{r}' \times (0, 0, -7.483) = 0$$

or

$$-M = \left(\frac{s'^2}{4} \times 0.6882 + \frac{s'}{2} \times 7.483 \times 0.6882 + 44.9, \right.$$

$$\left. \frac{s'^2}{4} \times 0.6882 + \frac{s'}{2} \times 7.483 \times 0.6882 + 29.9, 0 \right)$$

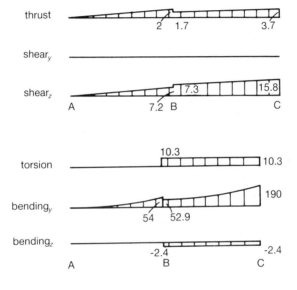

2.3.3.1 Differential Equations for Curved Beams

It is a relatively simple matter to derive the differential equations of equilibrium

$$\mathbf{P}' + \mathbf{p} = 0 \qquad\qquad (2.6)$$

$$\mathbf{M}' + \mathbf{R}' \times \mathbf{P} + \mathbf{m} = 0 \qquad\qquad (2.7)$$

for an arbitrary curved space beam in global coordinates. That is done in Fig. 2.18. The problem is that (a) these equations can be difficult to solve and (b) it is most common to work in the local coordinate system using standard stress resultant components.

The scalar form of these equations is written out fully in Example 2.20. Three special cases will be considered here:

Example 2.21 Equilibrium equations for curved beams.

Assume that the principal beam axes coincide with **n** and **b**. Write vector components in the local (intrinsic) coordinate system; for example,

$$\mathbf{P} = P_t\mathbf{t} + P_n\mathbf{n} + P_b\mathbf{b}, \text{ etc.}$$

$$\mathbf{P}' + \mathbf{p} = 0 \Rightarrow \frac{d}{ds}(P_t\mathbf{t} + P_n\mathbf{n} + P_b\mathbf{b}) + p_t\mathbf{t} + p_n\mathbf{n} + p_b\mathbf{b} = 0$$

Using the Frenet formulas (Appendix A)

$$P_t'\mathbf{t} + P_t\frac{\mathbf{n}}{\rho} + P_n'\mathbf{n} + P_n\left(\frac{\mathbf{b}}{\tau} - \frac{\mathbf{t}}{\rho}\right) + P_b'\mathbf{b} + P_b\left(\frac{\mathbf{n}}{\tau}\right) - p_t\mathbf{t} + p_n\mathbf{n} + p_b\mathbf{b} = 0$$

Three scalar equations:

$$P_t' - \frac{P_n}{\rho} \qquad + p_t = 0$$

$$P_n' + \frac{P_t}{\rho} - \frac{P_b}{\tau} + p_n = 0$$

$$P_b' + \frac{P_n}{\tau} \qquad + p_b = 0$$

Similarly for the moment equations

$$\mathbf{M}' + \mathbf{R}' \times \mathbf{P} + \mathbf{m} = 0 \Rightarrow M_t' - M_n/\rho \qquad + m_t = 0$$

$$M_n' + \frac{M_t}{\rho} - \frac{M_b}{\tau} - P_b \times m_n = 0$$

$$M_b' + M_n/\rho \qquad + P_n + m_b = 0$$

(using the facts that $\mathbf{R}' = \mathbf{t}$ and $\mathbf{t} \times \mathbf{P} = P_n\mathbf{b} - P_b\mathbf{n}$).

Special cases:

(1) Cable:

$$P_t' + p_t = 0$$
$$P_t/\rho + p_n = 0$$
$$p_b = 0$$

(2) Circular plane beam:

$$P_t' - \frac{P_n}{\rho} + p_t = 0$$

$$P_n' + \frac{P_t}{\rho} + p_n = 0$$

$$M_b' + P_n + m_b = 0$$

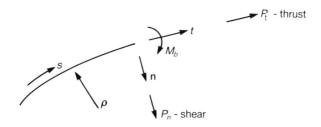

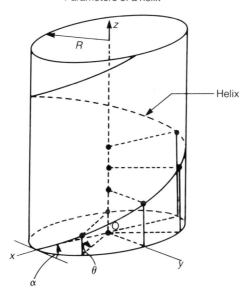

Parameters of a helix

---

(1) *The helix*. In this case all six equilibrium equations pertain; it is only necessary to note that for the helix $\rho = R/\cos^2 \alpha$ and $\tau = R/(\sin \alpha \cdot \cos \alpha)$. These follow directly from the fact that for the helix:

$$x = R \cos \theta$$
$$y = R \sin \theta$$
$$z = R\theta \sec \alpha$$

where R is the radius of the inscribed cylinder and α is the helix angle.

(2) *The cable*. For the case of a three-dimensional cable, the moment equations are trivially satisfied and there is only one non-zero force component P_t, i.e. $\mathbf{P} = P_t\mathbf{t}$.

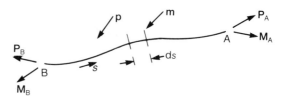

Curved beam

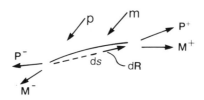

Typical element

$\mathbf{R}$ is the position vector, s is the arc length, $\mathbf{P}$, $\mathbf{m}$ are the internal forces and moments, $\mathbf{p}$, $\mathbf{m}$ are the applied forces and moments (distributed), and the prime indicates differentiation with respect to arc length, i.e. $\mathbf{R}' = d\mathbf{R}/ds$.

Sum forces:

$$d\mathbf{P} + \mathbf{p}\,ds = 0 \quad \text{or} \quad \boxed{\mathbf{P}' + \mathbf{p} = 0}$$

Sum moments:

$$d\mathbf{M} + d\mathbf{R} \times \mathbf{P} + \mathbf{m}\,ds = 0$$

(the moment of the distributed force $\mathbf{p}$ is a second order term in this equation), then

$$\boxed{\mathbf{M}' + \mathbf{R}' \times \mathbf{P} + \mathbf{m} = 0}$$

Fig. 2.18 – Equilibrium equations for a curved beam.

(3) *The circular plane beam.* In this case two force equilibrium equations remain together with the moment equilibrium equation about the **b** axis.

2.4 MEMBRANE SHELLS

While this text is primarily concerned with skeletal structures, there are analogous discussions for two- and three-dimensional continua. In particular there is a shell analog of the truss problem which is called membrane shell theory (see Fig. 2.19). It involves only in-plane shell forces and has no bending. This section will discuss the case of a symmetrically loaded spherical membrane shell which has common application in wood domes.

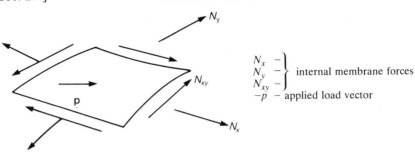

$$\left.\begin{array}{l} N_x \quad - \\ N_y \quad - \\ N_{xy} \quad - \end{array}\right\} \text{ internal membrane forces}$$

$-p$ – applied load vector

Fig. 2.19 – General membrane shell element.

2.4.1 A Symmetric Spherical Shell Under Its Own Weight

A particularly simple example of a membrane shell is the spherical dome under its own weight. Fig. 2.20 attempts to show this situation schematically. By assumption there are no bending moments shown on the shell

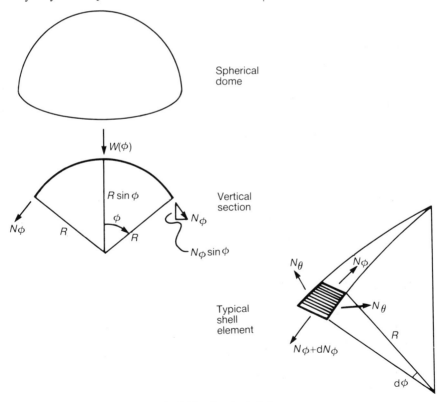

Fig. 2.20 – A spherical dome.

element; by symmetry there is no shear shown along element edges and no variation of N_θ with respect to the angle θ. The element must be in equilibrium under the action of the external load and the internal stress resultants N_ϕ and N_θ.

From arguments of summetry N_ϕ can be determined by the vertical equilibrium of a sector of the dome defined by the angle ϕ. Let $W(\phi)$ represent the total external vertical load downward on this sector. This load must then be carried by the vertical component of N_ϕ, $N_\phi \sin \phi$, integrated around the edge of the sector

$$W(\phi) + N_\phi \sin \phi \, 2\pi(R \sin \phi) = 0$$

or

$$N_\phi = \frac{-W}{2\pi R \sin^2 \phi}$$

When the symmetric vertical load $W(\phi)$ is the result of the weight q p.s.f. of a shell of uniform thickness, $W(\phi)$ can be computed as

$$W = \int_0^\phi 2\pi R^2 q \sin \theta \, d\theta = (1 - \cos \phi)2\pi R^2 q$$

Using this expression for W, N_ϕ then becomes

$$N_\phi = \frac{-Rq}{1 + \cos \phi}$$

The horizontal membrane force N_θ can now be computed using the horizontal equilibrium equation of the membrane shell element. (See Fig. 2.21.)

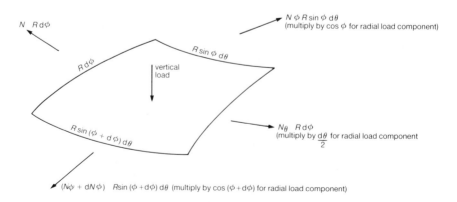

Fig. 2.21 – Spherical dome under a symmetric vertical load. Forces on the shell element.

$$NR \, d\phi \, d\theta + N_\phi \cos \phi R \sin \phi \, d\theta$$
$$- (N_\phi + dN_\phi)\cos(\phi + d\phi)R \sin(\phi + d\phi) \, d\theta = 0$$

or

$$N_\theta = \frac{d}{d\phi} (N_\phi \sin \phi \cos \phi)$$

For the particular case under discussion

$$N_\theta = -aq\left(\cos \phi - \frac{1}{1 + \cos \phi}\right)$$

2.5 EXERCISES

Analyze the plane trusses shown.

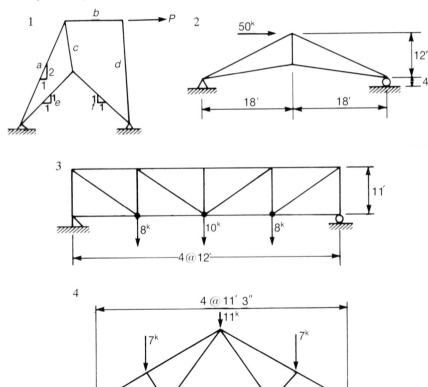

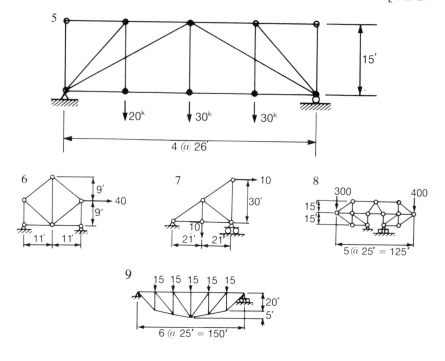

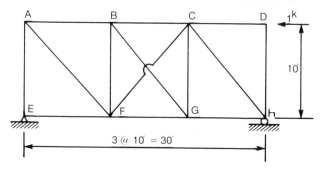

10 The force in bar BG = -1^k. What is the force in bar FC?

11 Show how the method of sections can be used to compute the forces in the diagonals of a Pratt truss.

12 Show how the method of sections can be used to compute the upper chord members in a Howe truss.

13 Add a symmetric load to the right span of Example 2.9 and resolve the Wichert truss.

14 Remove one of the applied loads in Example 2.12 and resolve the space truss. (Is the assumption that $R_6 = R_7$ still sensible?)

15 Resolve the Schwedler dome of Example 2.13 for the case in which only one joint is loaded.

16 Create a two-level Schwedler dome by stacking another layer (keep the bar slopes the same) on to the dome of Example 2.13. Apply a 'uniform load' to the upper level only and compute the bar forces.

17 The structure shown has full support at its periphery and is subjected to a unit downward load of its center. Assume bar force symmetry and compute all the bar forces.

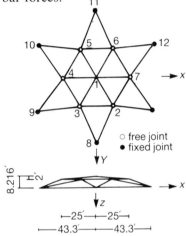

18 Assume bar force symmetry and analyze the truss shown.

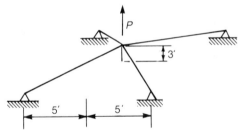

19 Solve the truss shown using joint equilibrium. Check your results using moment equations.

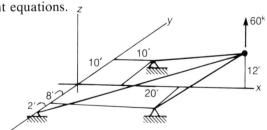

Compute and plot the stress resultants.

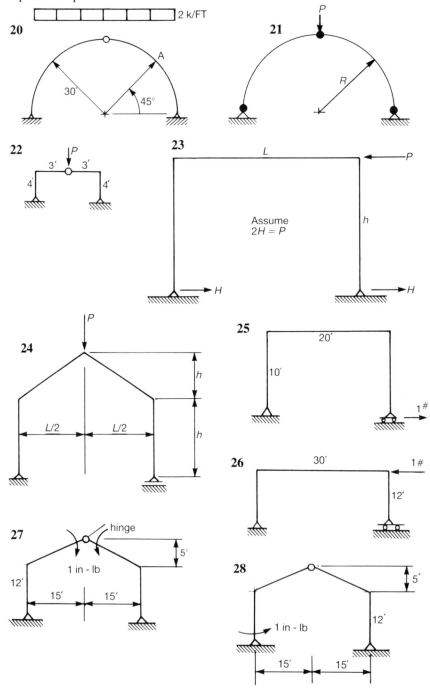

20

2 k/FT

30′ A 45°

21

P R

22

P 3′ 3′ 4′ 4′

23

L P

Assume
2H = P

h

H H

24

P L/2 L/2 h h

25

20′ 10′ 1#

26

30′ 1# 12′

27

hinge 1 in - lb 12′ 15′ 15′ 5′

28

1 in - lb 15′ 15′ 5′ 12′

29 For the semi-circular arch shown, compute and plot the moment, shear, and thrust diagrams.

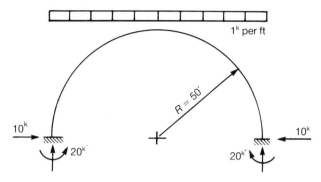

30 Solve the problem of a symmetric parabolic arch subjected to a single point load P at its crown.

31 Compute and plot stress resultants.

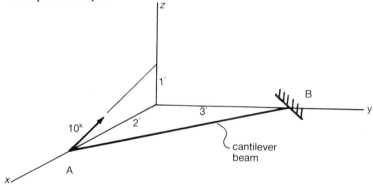

32 Compute and plot stress resultants.

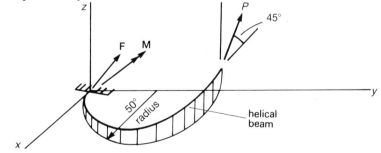

The beam shown is in the form of a helix with a radius of $50'$ and a helix angle of $30°$.

33 Compute and plot stress resultants.

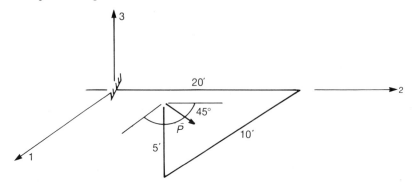

34 A coil spring (helical) is subjected to concentric loads. Compute the stress resultants by solving the differential equations directly. Check your results by using a free body diagram of a 'cut' spring.

35 Compute and plot the membrane solution for a symmetric spherical shell subjected to a point load of its peak.

36 For the helical beam of Example 1.5, compute and plot the stress resultants at four points equally spaced along the beam.

37 Compute and plot the shear, moment, and thrust diagrams.

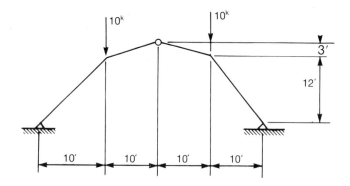

CHAPTER 3

Deflection of structures

This chapter presents the first material of the text to go beyond the concepts of elementary mechanics. When it is possible to remain within the realm of the equations of statics, structures differs little from basic mechanics. When, however, the discussion turns to the computation of displacements, a discussion of structures brings together some elements of statics and some elements of mechanics of solids, thus taking on more of a life of its own.

The technique used here to compute deflections is called the 'method of virtual work'. It is a powerful, classical method developed for problems of applied mechanics and requires only an inner product of equilibrium equations and a displacement like quantity (hence the designation virtual work). This is particularly important! The method of virtual work uses only equilibrium and does not require any assumptions concerning 'conservation of energy' for structural systems.

The principle of virtual work for a *rigid* body (see Chapter 1) states that 'the virtual work of a rigid body in equilibrium under a small rigid body motion is zero'. When applied to an *elastic* body in equilibrium the principle of virtual work will appear as 'the virtual work done by external loads equals the internal virtual strain energy'. There are many ways to approach such a theorem. It is common, for example, to assume or demonstrate that the principle of virtual work is valid for three-dimensional continua and then assume that it is valid for structures as a special case. That approach is in some ways devious since three-dimensional elasticity is usually beyond the background of undergraduate students. For simple structures such as trusses and plane frames it is possible to derive the virtual work principle directly. That will be done below. However in the extreme case of an arbitrarily curved three-dimensional beam, the equations themselves become more difficult and some authors assume a principle of virtual work in order to facilitate their derivation. This is simply by way of warning the reader to expect matters to become more complex as the chapter develops.

One final comment concerning the method of proof used in this chapter. While the virtual work theorems are conceptually simple and simple to demonstrate given a matrix formulation, without matrices their development tends to be cumbersome. For that reason, an inductive approach is used in the sections which follow. That is, while virtual work is first demonstrated for structural elements which are then combined to form a structure, the details of how these combinations actually occur are shown for examples of increasing complexity. For a more direct approach, the reader may wish to consult a text on matrix structural analysis.

3.1 VIRTUAL WORK FOR TRUSSES

This section is concerned with deriving the expression

$$\sum_{\text{nodes}} \mathbf{P}_i^{\text{v}} \cdot \boldsymbol{\delta}_i = \sum_{\text{members}} \Delta_i F_i^{\text{v}} \tag{3.1}$$

which will frequently take the form

$$1 \cdot \delta_1 = \sum_{\text{members}} \frac{F_i^{\text{v}} F_i L_i}{A_i E} \tag{3.2}$$

As do all applications of virtual work, these expressions involve two structural systems which may at any point become identical: First, there is a so-called 'virtual' system which must satisfy the equations of equilibrium and which bears the superscript 'v' above. Second, there is the so-called 'real' system for which displacements are to be computed. Eq. (3.2) then gives a desired displacement in terms of the bar forces F_i^{v} and F_i of the two systems, the length L_i and area A_i of each bar, and Young's modulus E. Several steps are required to derive this result.

3.1.1 Member Stiffness

There is a basic result of mechanics of solids which provides a realtionship between the force F_i in truss bar i and the length change of this bar Δ_i as

$$F_i = K_i \Delta_i \tag{3.3}$$

where F_i is the bar force in member i, Δ_i is the length change of member i, and K_i is the stiffness of member i.

In fact it will be shown that for straight uniform members K_i is simply

$$K_i = A_i E / L_i$$

where A_i is the cross-sectional area of member i, L_i is the length of member i, and E is Young's modulus.

This result follows simply from the fact that the strain ε_i for a truss bar is the length change Δ_i divided by the length L_i, or

$$\varepsilon_i = \Delta_i/L_i$$

It is assumed that the bar behaves in a linear elastic manner or that the stress σ_i and the strain ε_i are linearly related as

$$\sigma_i = E\varepsilon_i$$

Finally, the stress summed over the member area must equal the force on the cross-section,

$$F_i = A_i\sigma_i$$

and

$$F_i = A_i\sigma_i = A_iE\varepsilon_i = A_iE\Delta_i/L_i = K_i\Delta_i$$

as indicated above.

For the case of a non-uniform bar (see Fig. 3.1) whose area is a function of some spatial coordinate x it is convenient to start with a unit

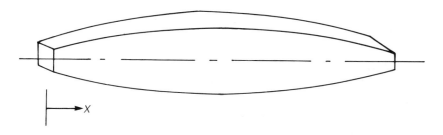

Fig. 3.1 – A non-uniform straight truss bar.

bar force F_i and note that K_i is then $1/\Delta_i$. The stress in this case is $1/A_i(x)$ and the strain is then $\varepsilon_i(x) = 1/(A_iE)$. The length change of the member can then be computed as

$$\Delta_i = \int_0^L (1/(A_iE)) \, dx$$

which is the reciprocal of the member stiffness as indicated above.

3.1.2 Member/Joint Displacement Relationship
This section will discuss briefly the commonly used linear relationship between member length change and the associated joint displacements (see Fig. 3.2)

$$\Delta_i = \mathbf{n}_i \cdot (\delta_A - \delta_c) \tag{3.4}$$

Equation (3.4) is an approximation as can be seen in Fig. 3.3 and is of course a projection of the joint displacements in the initial member direction.

The exact relationship between member length change and joint displacement can be written out easily but it is not linear. To use this non-linear relationship would require solving non-linear equations in order to

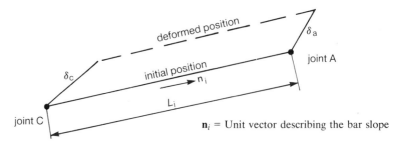

deformed position

δ_a

joint A

initial position

n_i

δ_c

L_i

joint C

n_i = Unit vector describing the bar slope

Fig. 3.2 – A typical bar.

perform structural analyses. Historically, linear analysis has proved adequate for the great majority of structures and thus has come to form the basis of introductory texts such as this book.

This is the first time any kind of analytical approximation has been described in this text. The fact of the matter is that the level of approximation involved in Eq. (3.4) is of the same order as the level of

Note: This is a special case in which only one end of a bar is allowed to displace. For the general case of Fig. 3.2 simply let $\delta \rightarrow \delta_A - \delta_C$.

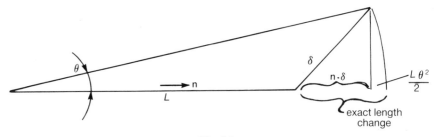

θ

δ

$n \cdot \delta$

$\frac{L\theta^2}{2}$

n

L

exact length change

Fig. 3.3

approximation involved when equilibrium equations are written in the undeformed configuration. This can be shown through the use of Taylor series expansions but has been omitted as beyond the scope of this text.

The justification given here will be simply that Eq. (3.4) is exact for the case of $\theta = 0$ (see Fig. 3.3) and apparently accurate for small angles $\theta \ll 1$. (In fact, much of mechanics is valid for cases of small rotations as is Eq. (3.4) above.)

3.1.3 Virtual Work for a Truss Element

Given a truss bar in equilibrium under a bar force F_i^v, it is possible to compute the virtual work of the associated end of member forces and a set of node displacements δ_A and δ_c (this condition is similar to that indicated in Fig. 3.2). Clearly

$$\mathbf{n}_i F_i^v \cdot \delta_A + (-\mathbf{n}_i F_i^v \cdot \delta_c) = F_i^v \mathbf{n}_i \cdot (\delta_A - \delta_c) = F_i^v \Delta_i \qquad (3.5)$$

using Eq. (3.4). Eq. (3.5) is equivalent to the statement that for a truss element the

virtual work of the node forces = the internal virtual strain-energy   (3.6)

Note that the node displacements are not related in any particular way to the member force at this point.

3.1.4 Virtual Work for Truss Structures

It only remains now to combine equations such as (3.5) into a virtual work equation for structures just as bars are combined to form structures.. Note first that as bars are combined to form structures, the internal virtual strain energy terms simply add to give the right-hand side of Eq. (3.1). Work of the node forces is somewhat more complex.

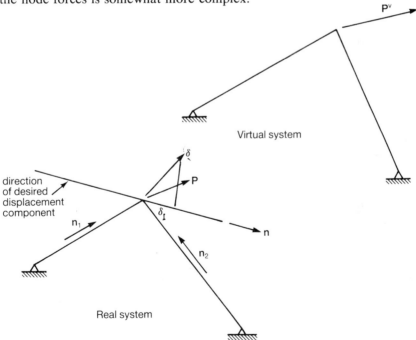

Fig. 3.4 – A simple truss.

As indicated in Fig. (3.4), if a truss *joint* is to be in equilibrium under the node load $\mathbf{P}_i$ *the end of member forces such as* $\mathbf{n}_\alpha \mathbf{F}_\alpha^v$ must sum to the applied load $\mathbf{P}_i^v$. It follows that the virtual work of the end of member forces must sum to

$$\sum_{\text{nodes}} \mathbf{P}_i^v \cdot \boldsymbol{\delta}_i$$

as indicated in the left-hand side of Eq. (3.1). Finally, if there is only a single unit load applied in the direction $\mathbf{n}$ in which it is desired to compute a displacement component (see Fig. 3.5), the sum

$$\sum_{\text{nodes}} \mathbf{P}_i^v \cdot \boldsymbol{\delta}_i \rightarrow \boldsymbol{\delta}_{\mathrm{I}}$$

as indicated in Eq. (3.2).

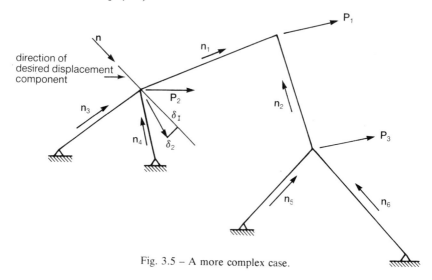

Fig. 3.5 – A more complex case.

3.1.5 Some Truss Examples

Figs. 3.4 and 3.5 are included here to indicate the manner in which terms combine as described above when bars are added to form a truss structure. For the system shown in Fig. 3.4, the virtual system is in equilibrium as

$$F_1^v \mathbf{n}_1 + F_2^v \mathbf{n}_2 = \mathbf{P}^v$$

Computing the dot product of this expression with the real joint displacement $\boldsymbol{\delta}$ results in

$$F_1^v \mathbf{n}_1 \cdot \boldsymbol{\delta} + F_2^v \mathbf{n}_2 \cdot \boldsymbol{\delta} = \mathbf{P}^v \cdot \boldsymbol{\delta}$$

or (using Eq. (3.4))

$$F_1^v \Delta_1 + F_2^v \Delta_2 = P^v \cdot \delta$$

Now if it is desired to compute the component δ_1 of δ in the direction implied by the unit vector $\mathbf{n}$ (i.e., $\delta_1 = \mathbf{n} \cdot \delta$) it is customary to take $\mathbf{P}^v = \mathbf{n}$ and obtain finally

$$F_1^v \Delta_1 + F_2^v \Delta_2 = \delta_1$$

which is a special case of the general formula indicated above in Eq. (3.2).

For the more complex configuration indicated in Fig. 3.5 the

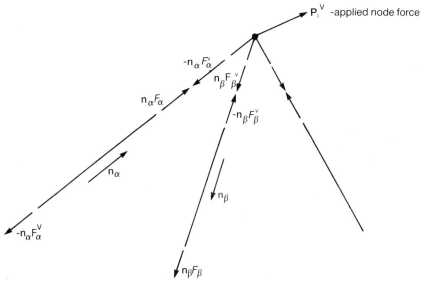

Fig. 3.6 – Typical node.

equilibrium equations of the virtual system are

$$F_1^v \mathbf{n}_1 + F_2^v \mathbf{n}_2 = \mathbf{P}_1^v$$
$$F_3^v \mathbf{n}_3 + F_4^v \mathbf{n}_4 - F_1^v \mathbf{n}_1 = \mathbf{P}_2^v$$
$$F_5^v \mathbf{n}_5 + F_6^v \mathbf{n}_6 - F_2^v \mathbf{n}_2 = \mathbf{P}_3^v$$

Again taking the dot product and combining terms gives

$$F_1^v(\delta_1 - \delta_2) \cdot \mathbf{n}_1 + F_2^v(\delta_1 - \delta_3) \cdot \mathbf{n}_2 + F_3^v \mathbf{n}_3 \cdot \delta_2 + F_4^v \mathbf{n}_4 \cdot \delta_2$$
$$+ F_5^v \mathbf{n}_5 \cdot \delta_3 + F_6^v \mathbf{n}_6 \cdot \delta_3 = \mathbf{P}_1^v \cdot \delta_1 + \mathbf{P}_2^v \cdot \alpha_2 + \mathbf{P}_3^v \cdot \delta_3$$

Again the left-hand side of this expression can be written as

$$F_1^v \Delta_1 + F_2^2 \Delta_2 + F_3^v \Delta_3 + F_4^v \Delta_4 + F_5^v \Delta_5 + F_6^v \Delta_6$$

If the joint loads in the virtual system degenerate to a specific unit load in a specific direction, the right-hand side of the above expression finally becomes simply the displacement component δ_1.

Having spent some time with the derivation of Eq. (3.2), its application is rather mechanical as indicated in Examples 3.1 and 3.2. Typically, a structure (the 'real' structure) is given and it is desired to compute a single scalar displacement component at some joint. Toward that end a second system (the virtual system) composed of the given structure and a single unit load corresponding to the desired displacement component is constructed. Both sets of bar forces must be computed. Finally, Eq. (3.2) must be evaluated; that is most conveniently done in tabular form.

This type of computation will be used repeatedly in Chapter 4 as part of the analysis of statically indeterminate structures.

Example 3.1 Problem: Compute the horizontal displacement of the upper left-hand joint of the truss shown in Fig. 2.9

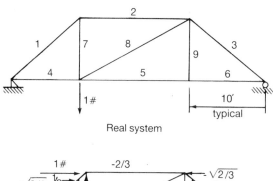

Real system

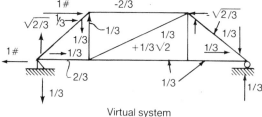

Virtual system

Bar	F	F^v	L	A	FF^vL/A
1	$-\frac{1}{3}\sqrt{2}$	$\sqrt{2}/3$	$10\sqrt{2}'$	2 in²	$-20\sqrt{2}/9$
2	$-\frac{2}{3}$	$-\frac{2}{3}$	10	2 in²	$20/9$
3	$-\sqrt{2}/3$	$-\sqrt{2}/3$	$10\sqrt{2}'$	2 in²	$10\sqrt{2}/9$
4	$\frac{2}{3}$	$\frac{2}{3}$	10	1	$40/9$
5	$\frac{1}{3}$	$\frac{1}{3}$	10	1	$10/9$
6	$\frac{1}{3}$	$\frac{1}{3}$	10	1	$10/9$
7	$\frac{2}{3}$	$-\frac{1}{3}$	10	1	$-20/9$
8	$\sqrt{2}/3$	$\sqrt{2}/3$	$10\sqrt{2}$	1	$10\sqrt{2}/9$
					74.14

$$\delta = \sum \frac{F_i^v F_i L_i}{A_i E} = \frac{74.14}{30 \times 10^6} = \underline{2.47 \times 10^{-6} \text{ ft}} \text{ . For steel } E = 30 \times 10^6 \text{ p.s.i.}$$

Example 3.2 Virtual work for Trusses
Given the 'real' structure and loads shown, find the vertical displacement of joint A.

$$\text{displacement} = \sum \frac{F_i F_i^v L_i}{A_i E} \qquad E = 29 \times 10^3 \text{ k.s.i. (steel)}$$

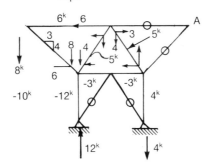

Real structure

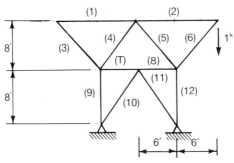

Virtual structure

Bar	F_i	F_i^y	L_i	A_i	$F_iF_i^yL_i/A_i$
1	6^k	0	12'	$\frac{1}{2}$ in^2	
2	0	3/4	12'	$\frac{1}{2}$	
3	-10^k	0	10'	$\frac{1}{2}$	
4	-5^k	5/8	10'	$\frac{1}{2}$	$-500/8$
5	5^k	$-5/8$	10'	$\frac{1}{2}$	$500/8$
6	0	$-10/8$	10'	$\frac{1}{2}$	
7	-3^k	$-3/8$	6'	1	$54/8$
8	-3^k	$-3/8$	6'	1	$54/8$
9	-12^k	$\frac{1}{2}$	8'	1	-48
10	0	0	10'	1	
11	0	0	10'	1	
12	4^k	$-1\frac{1}{2}$	8'	1	-48
					-1660
					8

$$\text{displacement} = -\frac{1660}{8 \times 29 \times 10^3} = \underline{-7.155 \times 10^{-3} \text{ ft}}$$

3.1.6. Williot Geometry

While all but forgotten today, graphical methods have in the past played an important role in structural analysis. One method, in particular, is extremely simple and yet gives insight into the type of approximations used above in computing joint displacements. It is called Williot geometry.

Williot geometry (Fig. 3.7) is based on the idea that if you wish to find the displaced position of a joint, you may simply think of rotating the deformed bars independently until they intersect. While this can be done

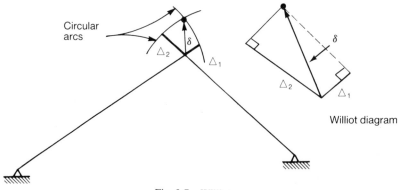

Fig. 3.7 – Williot geometry

well enough in the field at full scale, there are practical difficulties due to small length changes which arise when you try to do this on a small piece of paper to scale. Williot of course noted that for small angle changes you can approximate the circular arcs by their tangents and that then there is no point in drawing the entire bar. Only the displacements need be plotted. The diagram which remains is called a Williot diagram.

While useful as a graphical method, the Williot diagram can also be solved numerically as indicated in Example 3.3. That example also includes

---

Example 3.3   The use of Williot geometry.
Given $\Delta_1 = 0.1''$ and $\Delta_2 = 0.05''$ find the vertical joint displacement.

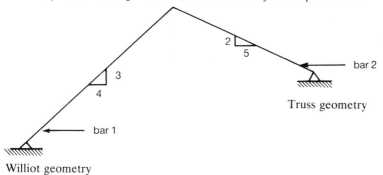

Truss geometry

Williot geometry

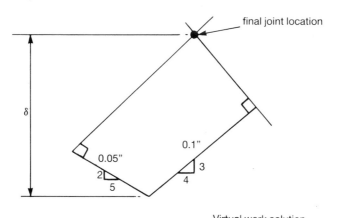

final joint location

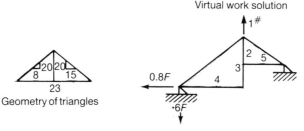

Geometry of triangles

Virtual work solution

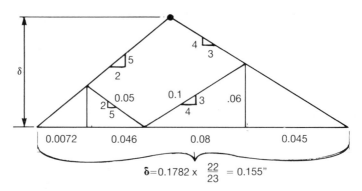

$$\delta = 0.1782 \times \frac{22}{23} = 0.155"$$

Virtual Work Solution

Moments about rt support

$$1 \times 5 + 0.8F \times 1 - 0.6F \times 9 = 0$$
$$F = 5/4.6 \quad \Rightarrow \quad 0.8F \qquad = 0.8695$$
$$\qquad = 1.087$$

$$\frac{5}{\sqrt{29}} G = 0.8695 \quad \Rightarrow \quad 6 \quad G = 0.936$$

$$\delta = 1.087 \times 0.1 + 0.936 \times 0.05$$
$$\quad = 0.155"$$

Check!

---

a virtual work solution and makes a point about virtual work. Computations of displacements in trusses can be geometrically complex. What virtual work really does for you is get rid of this complex geometry.

3.2 Virtual Work for Plane Frames

This section is concerned with deriving the expression

$$\sum_{\text{nodes}} (\mathbf{P}_i^v \cdot \boldsymbol{\delta}_i + T_i^v \theta_i) = \sum_{\text{members}} \left[\Delta_i F_i^v + \int_0^{L_i} (-y_i'') M_i^v \, dx \right] \qquad (3.7)$$

which will frequently take on the form

$$1 \cdot \delta_I = \sum_{\text{members}} \left[\frac{F_i F_i^v L_i}{A_i E} + \int_0^{L_i} \frac{M_i M_i^v \, dx}{EI_i} \right] \qquad (3.8)$$

These equations are clearly extensions of Eqs (3.1) and (3.2) above for trusses and it may simply be noted that $-y_i''$ is the curvature of member i in the real system, M_i^v is the moment of member i in the virtual system, I_i is the moment of inertia of member i, T_i is the applied moment of joint i, and θ_i is the rotation of joint i.

The derivation of Eqs (3.7) and (3.8) will follow steps similar to those used in the case of the truss problem.

3.2.1 The Moment–Curvature Relationship

In general, plane frames are composed of elements which are plane beams which of course may be curved. The discussion here, however, will be restricted to the case of *straight* beams for which the axial response (the 'truss effect') is uncoupled from the response to lateral load. In this case it is not necessary to repeat the above discussion concerning the member axial stiffness and attention can be turned to the question of the response of a beam to lateral load.

As is the case for trusses, in order to discuss beam displacements, it is first necessary to introduce a force/deformation or constitutive equation. For the plane beam this equation is

$$M = -EI\frac{d^2y}{dx^2} \tag{3.9}$$

Eq. (3.9) can be derived as follows (see Fig. 3.8). The strain at any point η on the cross-section is assumed to be a linear function of η. (This is called the assumption that 'plane sections remain plane'.)

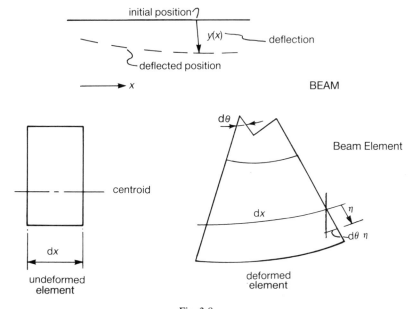

Fig. 3.8

$$\varepsilon = a + b\eta$$

For an elastic material stress is related to strain as $\sigma = E\varepsilon$. Now the stress integrated over the cross-section must be zero when no axial loads are present, that is

$$0 = \int_s \sigma \, dS$$

or

$$0 = \int_s E(a + b\eta) \, dS = Ea \times (\text{area}) + Eb \int_s \eta \, dS$$

when η is measured from the centroid of the cross-section, the last term in this equation is by definition zero which implies that $a = 0$ or that the stress is zero at the centroid of the cross-section.

Next, the stress distribution must be statically equivalent to the applied moment

$$M = \int_s \sigma\eta \, dS = \int_s Eb\eta^2 dS = EbI$$

where the moment of inertia I is defined as

$$I = \int_s \eta^2 \, dS$$

Finally, the strain can be written in terms of the curvature as

$$\varepsilon = \frac{d\theta\eta}{dx} \quad \text{and} \quad d\theta = -(dy') = -y'' \, dx$$

or

$$b\eta = -y''\eta \qquad b = -y''$$

and finally

$$M = EbI = -EIy''$$

In summary then the beam equations are

$$\frac{dV}{dx} = -w \qquad \frac{dM}{dx} = V \qquad M = -EI\frac{d^2y}{dx^2}$$

using the equilibrium equations of Chapter 2.

Having completed the derivation of the beam equations, any beam problem can now be solved. To illustrate the mechanics of doing so four specific examples are treated in Example 3.4. Note:

(a) The beam equations comprise a fourth order system

$$\left.\begin{array}{l} V' = -w \\ M' = V \\ M = -EIy'' \end{array}\right\} \quad \rightarrow \quad \left.\begin{array}{l} M'' = -w \end{array}\right\} \quad \rightarrow \quad (EIy'')'' = w$$

Example 3.4 Some beam problems.
1. *Fixed beam with uniform load*

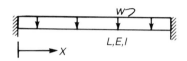

$$M'' = -w \Rightarrow M' = -wx + c_1 \Rightarrow M = -\frac{wx^2}{2} + c_1 x + c_2$$

but

$$M = -EIy''$$

$$-EIy'' = -\frac{wx^2}{2} + c_1 x + c_2$$

$$-EIy' = -\frac{wx^3}{6} + \frac{c_1 x^2}{2} + c_2 x + c_3$$

$$-EIy = -\frac{wx^4}{24} + \frac{c_1 x^3}{6} + \frac{c_2 x^2}{2} + c_3 x + c_4$$

Four boundary conditions:

$$y = 0 @ x = 0 \Rightarrow c_4 = 0$$

$$y' = 0 @ x = 0 \Rightarrow c_3 = 0$$

$$y' = 0 @ x = L \Rightarrow -\frac{wL^4}{24} + \frac{c_1 L^3}{6} + \frac{c_2 L^3}{2} = 0$$

$$y' = 0 @ x = L \Rightarrow -\frac{wL^3}{6} + \frac{c_1 L^2}{2} + c_2 L = 0$$

$$c_1 = \frac{wL}{2} \qquad c_2 = -\frac{wL^2}{12}$$

Final moment diagram:

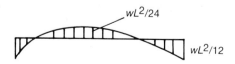

2. *Cantilever beam*

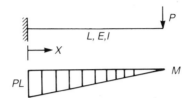

By inspection: $M = -P(L - x)$

$$-EIy'' = -P(L - x)$$

$$-EIy' = -PLx + \frac{Px^2}{2} + c_1$$

$$-EIy = -\frac{PLx^2}{2} + \frac{Px^3}{6} + c_1x + c_2$$

Two boundary conditions:

$$y = 0 \text{ @ } x = 0 \Rightarrow c_2 = 0$$
$$y' = 0 \text{ @ } x = 0 \Rightarrow c_1 = 0$$

3. *Fixed beam with concentrated load*

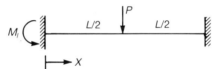

Using symmetry

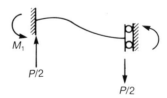

Let M_1 be the unknwon moment reaction

$$M = -M_1 + \frac{P}{2}x = -EIy''$$

$$-EIy'' = -M_1x + \frac{P}{2}\frac{x^2}{2} + c_1$$

$$-EIy = -\frac{M_1x^2}{2} + \frac{P}{2}\frac{x^3}{6} + c_1x + c_2$$

Three boundary conditions:

$$y = 0 \ @ \ x = 0 \Rightarrow c_2 = 0$$
$$y' = 0 \ @ \ x = 0 \Rightarrow c_1 = 0$$
$$y' = 0 \ @ \ x = \frac{L}{2} \Rightarrow M_1 = \frac{PL}{8}$$

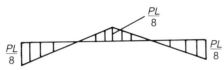

Final moment diagram

4. *Member stiffness*
Beam equation example

$$EIy'^v = w = 0$$
$$EIy''' = c_1$$
$$EIy'' = c_1 x + c_2$$
$$EIy' = \frac{c_1 x^2}{2} + c_2 x + c_3$$
$$EIy = \frac{c_1 x^3}{6} + \frac{c_2 x^2}{2} + c_3 x + c_4$$

Four boundary conditions:

$$y = 0 \ @ \ x = 0 \Rightarrow c_4 = 0$$
$$y' = 0 \ @ \ x = 0 \Rightarrow c_3 = 0$$
$$y = 0 \ @ \ x = L \Rightarrow \frac{c_1 L^3}{6} + \frac{c_2 L^2}{2} = 0$$
$$y' = 1 \ @ \ x = L \Rightarrow \frac{c_1 L^2}{2} + c_2 L = EI$$

the last two equations imply

$$\frac{c_1 L}{3} + c_2 = 0 \qquad \frac{c_1 L^2}{2} - \frac{c_1 L^3}{3} = EI$$

or

$$c_1 = \frac{6EI}{L^2} \qquad c_2 = -\frac{c_1 L}{3} = -\frac{2EI}{L}$$

$$M = -EIy'' = -c_1 x - c_2$$

$$@ \ x = 0 \qquad M = -c_2 \qquad = -\frac{2EI}{L}$$

$$@ \ x = L \qquad M = -c_1 L - c_2 = \frac{2EI}{L} - \frac{6EI}{L} = -\frac{4EI}{L}$$

Moment diagram:

It is therefore necessary to integrate four times to solve for the displacement given the load which implies that there must be four boundary conditions. Numbers 1 and 4 of Example 3.4 deal with this general case.

(b) Numbers 2 and 3 deal with special cases. For example, the cantilever is statically determinate so that it is possible to start with a known (or computed) moment diagram and integrate twice. The concentrated load problem uses symmetry to avoid the difficulties associated with matching together solutions at a discontinuous load.

In any case, these are all common beam problems and the solutions presented are all readily available in the literature.

3.2.2 The Virtual Work Equation for Beams

As stated above in the case of the truss, the virtual work expression is derived by taking the inner product of the equilibrium equations with a displacement like quantity and rearranging terms. For the case of the beam this involves starting with the equilibrium equation

$$\frac{d^2M}{dx^2} = -w$$

Multiplying by a displacement like quantity ξ and integrating over the length of the beam as

$$\int_0^L \frac{d^2M}{dx^2}\, \xi \, dx = -\int_0^L \xi w \, dx \qquad (3.10)$$

The left-hand side of this equation can now be integrated by parts as

$$\int_0^L \frac{d^2M}{dx^2}\, \xi \, dx = \xi \frac{dM}{dx}\Big|_0^L - \int_0^L \frac{dM}{dx}\frac{d\xi}{dx}\, dx$$

$$= \frac{dM}{dx}\Big|_0^L - M\frac{d\xi}{dx}\Big|_0^L + \int_0^L M \frac{d^2\xi}{dx^2}\, dx$$

Eq. (3.10) can finally be written as

$$\int_0^L w\xi \, dx = -\xi \frac{dM}{dx}\Big|_0^L + M \frac{d\xi}{dx}\Big|_0^L - \int_0^L M \frac{d^2\xi}{dx^2} \, dx \qquad (3.11)$$

Eq. (3.11) is the general virtual work expression for a beam. In terms of applications it is common to identify the equilibrium system (M, w, y) with the virtual loads and the variable ξ with the displacements in the real system. In order to have the left-hand side of Eq. (3.4) degenerate to the displacement at some point in the real beam, it is convenient to have the virtual load w degenerate to a concentrated unit load at some point $X = X_A$ as indicated schematically in Fig. 3.9. This figure shows a load of intensity

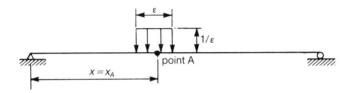

Fig. 3.9 – Concentrated load (see Appendix 3 for a more general discussion of discontinuous loads)

$1/\varepsilon$ is distributed uniformly over a length of ε while ε goes to zero. For this case the left-hand side of Eq. (3.4) becomes

$$\int_0^L w\xi \, dx = \int_{X_A - \varepsilon/2}^{X_A + \varepsilon/2} \xi \frac{1}{\varepsilon} \, dx = \xi_A \int_{X_A - \varepsilon/2}^{X_A + \varepsilon/2} \frac{1}{\varepsilon} \, dx = \xi_A$$

using the mean value theorem of calculus. Again the virtual work method implies the application of a unit load in the virtual structure to compute a displacement component at a point in the real structure. It will be noted here without proof that if it is desired to compute the rotation at some point X_A on a beam, a unit couple should be applied in the virtual system at this point. Using discontinuous functions properly, a distributed load w can be constructed to represent this case as was done for the case of the concentrated load. (See Appendix 3.)

Figure 3.10 shows an example of the use of the method of virtual work for computing beam deflections. In most applications the boundary terms are zero and the virtual work expression is used in the form

$$\xi_A = \int_0^L \frac{M^V M \, dx}{EI} \qquad (3.12)$$

3.2.3 A Note on the Integral of the Product of Two Functions

In computing the integrals used in the virtual work equation for beam deflections, the following result is sometimes useful.

THEOREM *When g is a linear function*

$$\int fg \, dx$$

can be written as the area under the curve of f times the value of the function g at the centroid of f.

Proof: $\int fg \, dx = \int f(a + bx) = a \int f \, dx + b \int fx \, dx$

$$= aA_f + bA_f \bar{x} = A_f(a + b\bar{x}) = A_f g \Big|_{x = \bar{x}}$$

where A_f is the area under the curve of f, and $\bar{x}$ is the centroid of the area under the curve of f.

Example 3.5 Problem: Find the center deflection of a uniformly loaded beam.

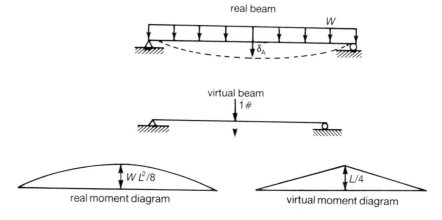

Using virtual work the deflection δ_A is

$$\delta_A = \int_0^L \frac{M^v M \, dx}{EI} = 2 \int_0^{L/2} \frac{M^v M \, dx}{EI} = 2 \times \frac{2}{3} \frac{L}{2} \frac{wL^2}{8} \times \frac{5}{8} \frac{L}{4} \frac{1}{EI}$$

$$= \frac{5}{384} \frac{wL^4}{EI}$$

Properties of half a parabola:

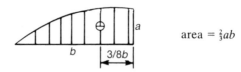

area $= \frac{2}{3}ab$

Problem: Find the rotation of the end of a cantilever subjected to a point load at its end.

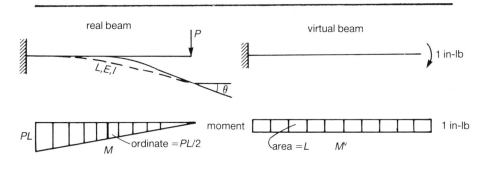

Using virtual work

$$\theta = \int_0^L \frac{M^v M \, dx}{EI} = \frac{1}{EI} \cdot L \cdot \frac{PL}{2} = \frac{PL^2}{2EI}$$

3.2.4 Virtual Work for Plane Frames

It remains to go from the virtual work expression for a beam to the virtual work expression for plane frames. In order to do so it is convenient to begin with a restricted form of Eq. (3.11) in which the distributed load w is zero. (This implies that displacements will only be computed in beams at node points but clearly a node can be added anywhere it is desired to compute a displacement.) In this case Eq. (3.11) becomes

$$\xi \frac{dM}{dx} \bigg|_0^L - M \frac{d\xi}{dx} \bigg|_0^L = - \int_0^L M \frac{d^2\xi}{dx^2} \, dx \qquad (3.13)$$

This equation again demonstrates the familiar form

virtual work of the node forces = internal virtual strain energy   (3.14)

which can be seen by noting that since $dM/dx = V$ the first term of Eq. (3.13) represents the work done by the shearing forces at each end of the beam; the second term represents the work done by the bending moments at each end. (The sign of this term is due to the fact that a positive moment corresponds to negative curvature.)

The earlier argument concerning the end of member forces summing to the joint load for trusses still holds but it must be extended to include moment equilibrium. The result is Eq. (3.7) which degenerates to Eq. (3.8) when only a single virtual load is applied.

The example of Fig. 3.10 will now be used to show exactly how terms combine in a specific case of the virtual work expression for plane frames. As stated above, the joint equilibrium equations are first written as

$$\textit{Joint 1} \qquad\qquad \textit{Joint 2}$$

$$
\begin{aligned}
t_2 - V_{11} &= 0 & -t_2 + V_{32} &= 0 \\
-t_1 - V_{21} &= 0 & V_{22} - t_3 &= 0 \\
M_{21} + M_{11} &= 0 & M_{22} - M_{32} &= 0
\end{aligned}
$$

Multiplying these equations by the joint displacement components in the real structure δ_{1x}, δ_{1y}, θ_1, δ_{2x}, δ_{2y}, θ_2 and forming the sum

$$
\begin{aligned}
(t_2 - V_{11})\,\delta_{1x} &- (t_1 + V_{21})\,\delta_{1y} + (M_{21} + M_{11})\,\theta_1 \\
&+ (-t_2 + V_{32})\,\delta_{2x} + (V_{22 - t_3})\,\delta_{2y} + (M_{22} - M_{32})\,\theta_2 = 0
\end{aligned}
$$

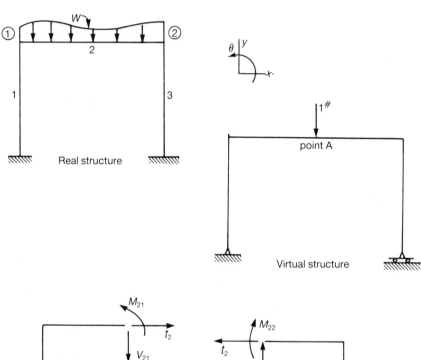

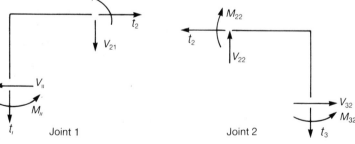

Fig. 3.10

These terms can now be combined as

$$t_2\,\delta_{1x} - t_2\,\delta_{2x} = -t_2^V\,\Delta_2$$

$$-t\,\delta_{1y} \qquad\quad = -t_1^V\,\Delta_1$$

$$-t\,\Delta_{2y} \qquad\quad = -t_3^V\,\Delta_3$$

$$-V_{21}\,\Delta_{1y} + M_{21}\,\theta_1 + V_{22}\,\delta_{2y} - M_{22}\,\eta_2 = \int_0^L w_2\,\xi_2\,dx_2 + \int_0^L M_2^V\,\frac{d^2\xi_2}{dx_2^2}\,dx_2$$

$$-V_{11}\,\delta_{1x} + M_{11}\,\theta_1 = \int_0^L M_1^V\,\frac{d^2\xi_1}{dx_1^2}\,dx_1$$

$$V_{32}\,\delta_{2x} + M_{32}\,\theta_2 = \int_0^L M_2^V\,\frac{d^2\xi_3}{dx_3^2}\,dx_3$$

giving the virtual work expression

$$\int_0^L {}_2\xi_2\,dx_2 = t_1^V\,\Delta_1 + t_2^V\,\Delta_2 + t_3^V\,\Delta_3 - \int_0^{L_1} M_1^V\,\frac{d^2\xi_1}{dx_1^3}\,dx_1$$
$$-\int_0^{L_2} M_1^V\,\frac{d^2\xi_2}{dx_2^2}\,dx_2 - \int_0^{L_3} M_3^V\,\frac{d^2\xi_3}{dx_3^2}\,dx_3$$

as indicated earlier.

3.2.4.1 Applications

As in the case of the truss and in spite of all the time spent discussing its derivation, the virtual work method for frames is rather mechanical in its application (see Example 3.6). It is first necessary to perform two frame analyses, one for the real structure and one for the virtual structure whose loading is dictated by the displacement component desired. It is then necessary to evaluate the virtual work expression which contains one truss-like term per bar and one integral per bar over the product of the real and virtual moment diamgrams. Since the effort required to do this is beginning to mount it is important to proceed effectively. Toward that end, the little theorem on integrals of products from this chapter can be useful.

The first case of Example 3.6 is a three-hinged arch-type frame acted upon by a uniform loading. Since the structure is symmetric it is only necessary to deal with half of the members and then multiply their contribution by a factor of two. A special property of parabolas is indicated. The second case is a simple statically determinate plane frame.

A final comment on the matter of units. While displacements and rotations are being computed, the virtual work expressions, of course, carry the units of work. The result must then be thought of as the work

Example 3.6 Virtual work for frames.

Case 1

Given: $E = 30 \times 10^3$ k.s.i.

$\left.\begin{array}{l} I = 20\,000 \text{ in}^4 \\ A = 10 \text{ in}^2 \end{array}\right\}$ both members

Find the vertical deflection of point A.

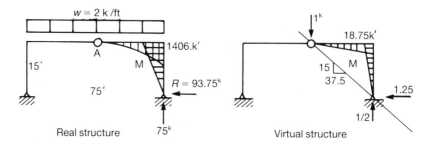

Real structure   Virtual structure

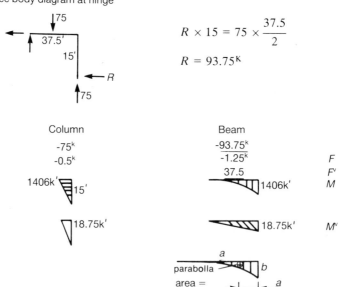

Free body diagram at hinge

$$R \times 15 = 75 \times \frac{37.5}{2}$$

$$R = 93.75^{\text{K}}$$

Virtual work:

$$\delta = \sum_{\text{members}} \left[\frac{F_i F_i^y L_i}{A_i E} + \int \frac{M_i M_i^y \, dx}{EI} \right]$$

$$\delta = \frac{2 \text{ in}^2}{30 \times 10^3 \text{ K}} \left\{ \frac{75 \times \frac{1}{2} \times 15 + 93.75 \times 1.25 \times 37.5}{10} \frac{12 \text{ in}}{FT} \right.$$

$$\left. + \frac{\frac{1}{3} \times 1406 \times 18.75 + \frac{1}{3} \times 37.5 \times 1406 \times \frac{3}{4} \times 18.75}{20\ 000} \frac{12^3 \text{ in}^3}{FT^3} \right\}$$

$$= \frac{2}{15 \times 10^3} (5948 + 32{,}742) = \underline{2.58 \text{ in}}$$

Case 2

Given the loaded frame find the vertical displacement at point A

$$E = 30 \times 10^3 \text{ k.s.i.}$$
$$I = 15\ 000 \text{ in}^4$$
$$A = 8 \text{ in}^2$$

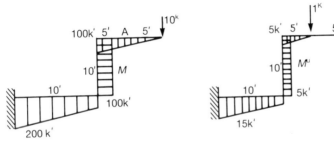

Real structure   Virtual structure

Member	1	2	3
F	0	-10^K	0
F^v	0	-1^K	0
M			
M^v			

Virtual work:

$$\delta = \sum_{\text{members}} \left[\frac{F_i F_i^v L_i}{A_i E} + \int \frac{M_i M_i^v \, dx}{EI_i} \right]$$

$$\delta = \frac{1}{30 \times 10^3} \left\{ \frac{10 \times 10}{8} \times 12 \right.$$

$$+ \frac{\frac{1}{2} \times 5 \times 5 \times 83.3 + 5 \times 10 \times 100 + 100 \times 10 \times 10 + 100 \times 10 \times \frac{1}{2} \times 11.67}{15\,000} \times 12^3 \left.\right\}$$

$$= \frac{1}{30 \times 10^3} (150 + 2520) = \underline{0.089 \text{ in}}$$

done by a unit force or the work done by a unit couple which is equal numerically to the desired displacement or rotation. The reader should also be particularly careful about units when combining terms of bending and axial load.

3.3 MORE GENERAL APPLICATIONS

There is an entire spectrum of applications of virtual work ranging upward from elementary beam theory to arbitrarily curved beams in space and including somewhat peripheral topics such as shear deformation, warping of cross-sections under torsion . . . Happily, many of these topics can be argued on the basis of the discussion already presented:

(1) *Curved plane beams*. It is common practice, for example in arch bridges, to apply to curved members the virtual work expression from Section 3.3 which was derived for straight members. The problem with curved beams is that under the assumption that plane sections remain plane the stress distribution is no longer linear. It is argued in this case that for the large radii of curvature commonly encountered in arches, this non-linearity is small and can be neglected.

(2) *Straight three-dimensional beams*. Problems of straight three-dimensional beams without the warping of cross-sections in torsion reduce to bending about two axes plus uncoupled axial load and torsion. Since these are at worst plane problems, serious three-dimensional complications do not arise. In this case the virtual work expressions can be written as

$$\delta = \sum_{\text{members}} \frac{t_i t_i^V L_i}{A_i E} + \sum_{\text{members}} \int_0^{L_i} \frac{M_{ix} M_{ix}^V \, dx_i}{EI_{ix}}$$

$$+ \sum_{\text{members}} \int_0^{L_i} \frac{M_{iy} M_{iy}^V \, dx_i}{EI_{iy}} + \sum_{\text{members}} \int_0^{L_i} \frac{T_i T_i^V \, dx_i}{J_i G}$$

The two additional terms in this expression reflect the fact that space frames have torsion and bending in addition to the effects present in plane frames. In the above expression each bending moment M_{ix}, M_{iy} has its own associated moment of inertia I_{ix}, I_{iy}; the torsion T_i has its associated stiffness J_i and G is the shear modulus.

(3) *The general case of beams in three dimensions.* Appendix 5 moves toward completing a three-dimensional theory of flexure by deriving the appropriate strain–curvature expressions. Leaving matters there it will simply be noted that using earlier arguments, the virtual work expression derived above for straight three-dimensional beams can be applied to some common three-dimensional arch problems again when the radii of curvature remain large.

3.4 COMPUTATION OF DISCONTINUITIES

Many applications in this text will use virtual work to compute discontinuities associated with releases which have been introduced in members. In these cases two virtual forces will be applied as indicated in Fig. 3.11. In

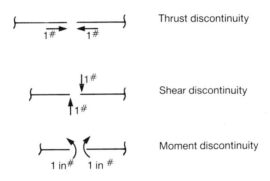

Fig. 3.11

all cases the logic is the same and only the thrust discontinuity will be discussed.

A unit virtual load applied to the end of a truss bar would measure the deflection of the bar in the direction of the force. The opposing unit forces then measure the amount that the two ends move together which is the definition of the axial discontinuity.

3.5 REAL STRUCTURES versus VIRTUAL STRUCTURES

By way of extending the results of this section it will be noted here that the real structure and the virtual structure need not be identical. In fact, all that is required of the virtual structure is an equilibrium solution. Arguments are sometimes made in this case for subjecting the virtual structure to the deformations of the real structure. The reader may verify for himself that the above discussions of virtual work are valid in this case.

In practical terms this means for trusses or frames a statically determinate virtual substructure may be used given a statically indeterminate real structure. Two examples are included here to make this point. Example 3.7 discusses the problem of computing the central deflection of a fixed beam. Note that the virtual structure is simply supported. Example

---

Example 3.7 Virtual structure different from the real structure. Find the center deflection of a fixed ended beam with a point load P at its center.

$$\delta = \int \frac{M M^v \, dx}{EI} = \frac{2}{EI} \left(\frac{PL}{4} \frac{L}{2} \frac{1}{2} \times \frac{2}{3} \frac{L}{4} - \frac{PL}{8} \frac{L}{2} \frac{L}{8} \right) = \frac{PL^3}{EI} \frac{1}{16 \times 12}$$

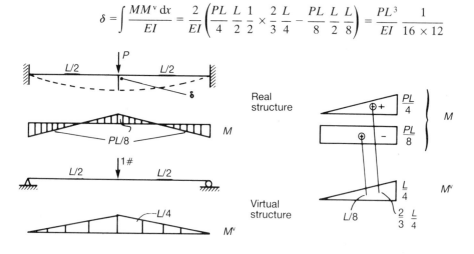

Example 3.8 Virtual structure different from real structure

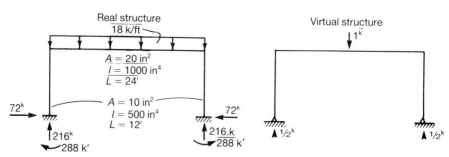

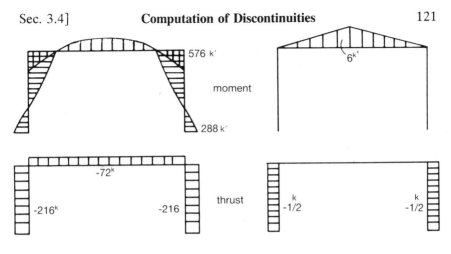

$$\delta = \sum_{\text{members}} \frac{t_i t_i^{\text{v}} L_i}{A_i E} + \sum_{\text{members}} \int_0^{L_i} \frac{M_i M_i^{\text{v}} \, \mathrm{d}x}{EI_i}$$

Using symmetry there is only one axial load term and one bending term to be evaluated.

Axial load term

$$\frac{-1/2 \times -162 \times 12}{10 \times 30 \times 10^3} = 0.00324 \text{ ft}$$

Bending term ($wL^2/8 = 1298$ K^1)

$$(\tfrac{2}{3} \times 1298 \times 12 \times \tfrac{5}{8} \times 6 - \tfrac{1}{2} \times 576 \times 12 \times 2) \times \frac{144}{30 \times 10^3 \times 100} = 0.1538 \text{ ft}$$

Summing terms and multiplying by 2 (for symmetry) gives the final displacement to be

$$= 2 \times (0.0032 + 0.1538) = 0.314 \text{ ft}$$

3.8 discusses a similar frame problem. In both cases, if the real structure were to be used as the virtual structure an indeterminate analysis would be required to determine the virtual forces. Using a statically determinate real structure then avoids that analysis and simplifies matters.

3.6 EXERCISES

1 Compute the vertical deflection at point A.

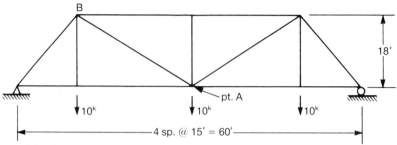

Members areas:

        Web members   4 in^2
        upper chords   5 in^2
        lower chords   3 in^2

$E = 29 \times 10^6$ p.s.i. (steel)

2 Compute the center deflection of a beam acted upon by a concentrated load at its center.

3 Compute the horizontal deflection of point A. Assume E, I, A to be constant.

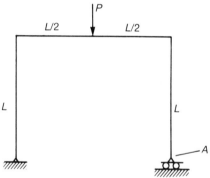

4 Camber of trusses. There is a physiological phenomenon through which the eye will see a vertical retaining wall as leaning outward and a horizontal truss as sagging. In order to avoid this effect and also to counteract dead load deflections a truss may be cambered, i.e. built a little above horizontal. Virtual work can easily be used to deal with computations required in this case.

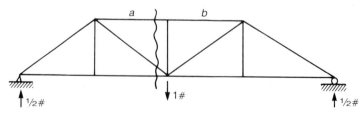

For example, suppose it is desired to camber the truss of Exercise 1 6 in by fabricating the two upper chord members too long by an amount ΔL each. In this case the bar forces $F_a = F_b = -15/18$. The most convenient form of virtual work is

$$\delta = \sum_{\text{members}} F_i^v \, \Delta L_i$$

$$-6 = 2 \times \left(-\frac{15}{18}\right) \Delta L \Rightarrow \Delta L = 3 \times \tfrac{6}{5} \text{ in.}$$

Resolve this problem for the case of a five panel truss.

5 Find $y(x)$ $0 < x < L$ using the method of virtual work.

6 Find the rotation of the beam at its support using the method of virtual work.

7 Solve this problem of support displacement by integrating the differential equation.

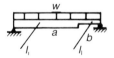

8 Solve the variable moment of inertia problem shown.

9 Use the variation of moment of inertia given in Exercise 8 and find M as a function of θ.

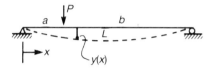

10 Assume all bars have the same area A and solve for the vertical displacement of the Schwedler dome of Example 2.13 under a 'uniform load'.

11 Assume all bars have the same area A and solve for the displacement of the loaded points of Example 2.12.

12 Compute the horizontal displacement of point B in Exercise 1 above.

13 It is common in some cases of rigid frames and arches to neglect truss-type terms ('axial load effects'). Do so and compute the deflection of a semicircular three-hinged arch under a point load at its center.

14 Find the value of d indicated. Assume $I = 1000$ in^4 $A = 5$ in^2 $E = 30 \times 10^6$ p.s.i. (steel).

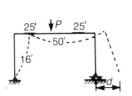

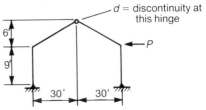

15 The frame of Example 3.8 is cut at its center. Compute the three discontinuities which develop at this cut.

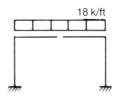

16 Compute the vertical deflection of the free end of the beam of Example 2.20. Assume $E = 3 \times 10^6$ p.s.i. (concrete) and a $5' \times 8\frac{1}{4}''$ slab.

17 Discuss the stiffness of an automobile coil spring as a function of its parameters.

18 Find the vertical deflection (in inches) of point A using the method of virtual work. Assume

$I = 2000$ in^4
$A = 10$ in^2
$E = 30 \times 10^6$ p.s.i.

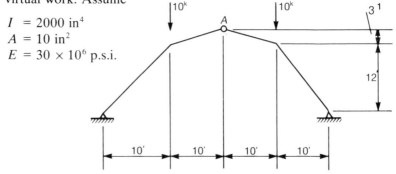

Solution to Exercise 18

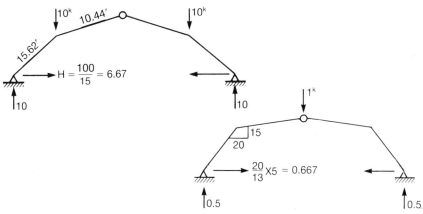

Lower member

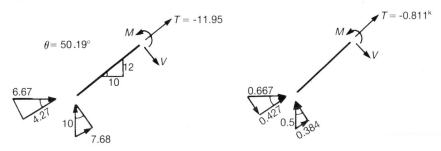

Upper member

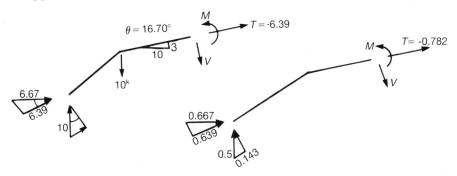

Moment at kink:
$$M = 10 \times 10 - 6.67 \times 12 = 20^{k'}$$

$$M = 0.5 \times 10 - 0.667 \times 12 = -3^{k'}$$

$$displacement = \sum_{\text{members}} \left[\frac{F_i F_i^y L_i}{A_i E} + \int \frac{M_i M_i^y \, dx}{EI} \right]$$

$$= \frac{2}{E} \left\{ \frac{11.95 \times 0.81 \times 15.62}{10} + \frac{6.39 \times 0.782 \times 10.44}{10} - \frac{1}{3} \right.$$

$$\left. \times \frac{144 \times 20 \times 3}{2000} \times (15.62 + 10.44) \right\}$$

$$= \frac{2}{E} \{20.35 - 37.53\} = - \frac{17.7 \times 2}{30 \times 10^3}$$

$$= -1.14 \times 10^3 \, ft = -13.7 \times 10^{-3} \, in.$$

CHAPTER 4

Statically indeterminate structures

Fig. 4.1 shows two extremely simple structures. In both cases it is possible to take either of the bar forces to be arbitrary and still satisfy the equations of equilibrium. The equilibrium equations thus have many solutions and for that reason both structures are said to be statically indeterminate. (See Chapter 2.)

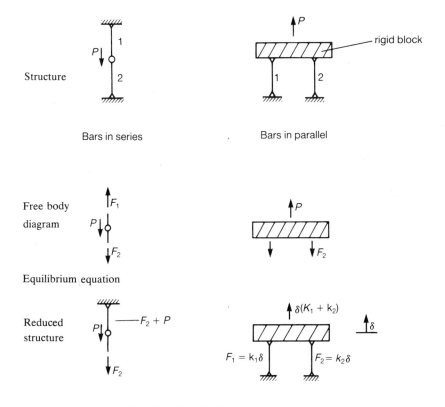

Fig. 4.1 – Two simple indeterminate structures.

Statically indeterminate structures are a principle topic of the theory of structures and the subject of this chapter. The fact that the equilibrium equations in this case have no unique solution is annoying to the extent that a given load should produce a unique response from a linear system. The problem is, of course, that the equations of statics do not completely *define* a structure; to complete the description it is necessary to introduce member stiffnesses into the formulation. Once that has been done the expected unique system response follows directly.

The role of member stiffness is obvious in the two cases shown in Fig. 4.1. Intuitively, stiffer members carry more of the load. Put another way, as the area of bar 2 goes to zero in each case bar 1 is forced to carry the entire load P; the argument is symmetric to the extent that as bar 1 goes to zero bar 2 must carry the load. There are thus simple bounds on the manner in which these systems can respond.

What eliminates the lack of uniqueness found in the equilibrium equations is the fact that 'the pieces of a deformed structure must fit together'. That fact will be enforced here in two different ways and thus give rise to two different methods of analysis, a force method and a displacement method. For example, if the lower support is removed in the case of the two bars in series, the structure becomes statically determinate and the displacement at this support can be computed by summing the length changes of both bars

$$\Delta_1 + \Delta_2 = \frac{F_2 + P}{K_1} + \frac{F_2}{K_2}$$

where

$$k_1 = A_1 E / L_1$$
$$k_2 = A_2 E / L_2$$

Since the support displacement must be zero it follows that

$$\Delta_1 + \Delta_2 = 0 \Rightarrow F_2 = \frac{-k_2 P}{k_1 + k_2} \Rightarrow F_1 = F_2 - P = \frac{k_1 P}{k_1 + k_2}$$

Since a force (F_2) is used as the unknown here this type of analysis is called the 'force method'.

The parallel bars of Fig. 4.1 are an obvious candidate for another type of analysis. In this case it is clear that if the displacement δ is known the bar forces can be computed as

$$F_1 = k_1 \delta \quad \text{and} \quad F_2 = k_2 \delta$$

which implies an applied load of

$$(k_1 + k_2) \delta$$

But since the applied load must be P, it follows that

$$P = (k_1 + k_2) \delta \Rightarrow \delta = \frac{P}{k_1 + k_2}$$

and

$$F_1 = K_1 \delta = \frac{k_1 P}{k_1 + k_2}, F_2 = k_2 \delta = \frac{k_2 P}{k_1 + k_2}$$

Since a displacement is used as the unknown in this analysis it is called a 'displacement method'.

This chapter is concerned with formalizing the force and displacement methods so that they can be applied to arbitrary structures. The methods are related to the extent that they both begin by introducing procedures which reduce the structure to some case which can be dealt with easily (such as a statically determinate structure). This reduction will be shown to violate the definition of the given structure. Finally a solution is constructed to repair these violations.

These introductory comments attempt to show that the logic of the analysis of statically indeterminate structures is simple. The fact that these methods sometimes appear complex in application is due to the inherent geometrical complexity of structures which surfaces when displacements or forces must be computed.

4.1 THE FORCE METHOD

Given the 'idea' of the force method just presented, this section develops a systematic approach with three steps. It begins with the discussion of a second single degree of freedom structure which is more practical than the two-bar truss discussed. It then moves on to a two-degree of freedom system which finally leads to a general statement of the force method.

4.1.1 A Single Degree of Freedom System
The force method can easily be described as a sequence of steps:

Step 1. Reduce the structure to a statically determinate structure. That is done in Example 4.1 by replacing the right-hand horizontal support by a

---

Example 4.1   Example of the force method.

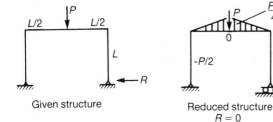

Given structure

Reduced structure
R = 0

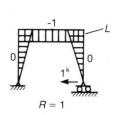

R = 1

$$\delta = \sum_{\text{members}} \left[\frac{F_i F_i^V L_i}{A_i E} + \int \frac{M_i M_i^V \, dx}{EI} \right]$$

Deflection at R due to the load P

$$\delta_0 = - \frac{1}{EI} \left[\frac{L}{2} \cdot \frac{PL}{4} L \right] = - \frac{PL^3}{8EI}$$

Deflection due to unit R

$$\delta_1 = \frac{L}{AE} + \frac{2}{EI} \left[L \cdot L \cdot \frac{L}{2} + \frac{1}{3} L^2 L \right] = \frac{L}{AE} + \frac{10}{6} \frac{L^3}{EI}$$

$$\delta_0 + R \, \delta_1 = 0 \Rightarrow \boxed{R = - \frac{\delta_0}{\delta_1} = \frac{PL^3}{8EI} \frac{1}{\dfrac{L}{AE} + \dfrac{10}{6} \dfrac{L^3}{EI}}}$$

Final forces and moments

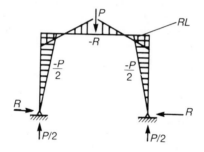

roller. This step allows the structure to displace where it was formerly fixed. The reaction R which is temporarily set to zero can be used to define a displacement (discontinuity) δ and its direction.

Step 2. Compute the value of δ due to the load. The method of virtual work is used to compute the displacement of the structure at its right support. In this case the reduced structure is the real structure and the structure marked $R = 1$ is the virtual structure. Since this support moves to the right while R and δ have been taken as positive to the left, δ_0 comes out to be negative.

Step 3. Compute the value of δ due to a unit value of R. Again, virtual work is used to compute the motion at the support. In this case the structure marked $R = 1$ serves as both the virtual and real structures.

Step 4. Solve for the reaction R. R is of course the value of the reaction which is required in order to 'push the structure back into place'.

Step 5. Compute and plot the final stress resultant diagrams.

4.1.2 A Two-Degree of Freedom System

The trouble with the single degree of freedom system is that there is an interaction between redundants which it does not demonstrate. Exanple 4.2 shows this interaction clearly. It is again convenient to proceed in steps:

Step 1. Reduce the structure to a statically determinate structure. In the case of Example 4.2 this requires that two 'cuts' be made and thus defining two redundants. (These redundants are taken to be positive when they place their respective bars in tension.)

Step 2. Analyze the structure. It will be necessary below to have the bar forces for three cases of load: the structure under the given load and the structure under unit values of the redundants R_1 and R_2.

Example 4.2

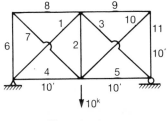

Given structure

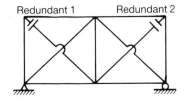

'Cut' Structure

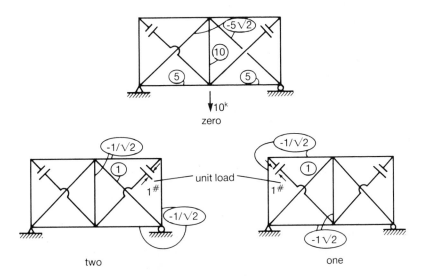

Bar	L	A	F_0	F_1	F_2	$\dfrac{F_0F_1L}{A}$	$\dfrac{F_0F_2L}{A}$	$\dfrac{F_1^2L}{A}$	$\dfrac{F_2^2L}{A}$	$\dfrac{F_1F_2L}{A}$
1	$10\sqrt{2}'$	3 in^2	$-5\sqrt{2}\,K$	1	–	$-\dfrac{100}{3}$		$10\sqrt{2}/3$		
2	$10'$	1 in^2	$10\,K$	$-\dfrac{1}{\sqrt{2}}$	$-\dfrac{1}{\sqrt{2}}$	$-\dfrac{100}{\sqrt{2}}$	$-\dfrac{100}{\sqrt{2}}$	5	5	5
3	$10\sqrt{2}'$	3 in^2	$-5\sqrt{2}\,K$		1		$-\dfrac{110}{3}$		$10\sqrt{2}/3$	
4	$10'$	2 in^2	$5\,K$	$-\dfrac{1}{\sqrt{2}}$		$-\dfrac{25}{\sqrt{2}}$		$2\tfrac{1}{2}$		
5	$10'$	2 in^2	$5\,K$		$-\dfrac{1}{\sqrt{2}}$		$-\dfrac{25}{\sqrt{2}}$		$2\tfrac{1}{2}$	
6	$10'$	1 in^2		$-\dfrac{1}{\sqrt{2}}$				5		
7	$10\sqrt{2}$	3 in^2		1				$10\sqrt{2}/3$		
8	10	2 in^2		$-\dfrac{1}{\sqrt{2}}$				$2\tfrac{1}{2}$		
9	$10'$	2 in^2			$-\dfrac{1}{\sqrt{2}}$				$2\tfrac{1}{2}$	
10	$10\sqrt{2}$	3 in^2			1				$10\sqrt{2}/3$	
11	$10'$	1 in^2			$-\dfrac{1}{\sqrt{2}}$				5	
						-121.6	-121.6	24.42	24.42	5
						(δ_{10})	(δ_{20})	(δ_{11})	(δ_{22})	$(\delta_{12}=\delta_{21})$

Superposition:

final solution = zero solution + $R_1 \times$ solution one + $R_2 \times$ solution two

Results:

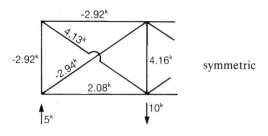

symmetric

Step 3. *Set up the equations of superposition and compute the coefficients.* Two simultaneous equations will be used to determine the bar forces R_1 and R_2,

$$\delta_{11}R_1 + \delta_{12}R_2 + \delta_{10} = 0$$
$$\delta_{21}R_1 + \delta_{22}R_2 + \delta_{20} = 0$$

(4.1)

This is now the heart of the force method. Physically it is required to select R_1 and R_2 so that the ends of the cut bars are not separated in the final solution. ('The pieces must fit together.') The interaction mentioned above arises because a change in R_1 causes the ends of bar 2 to separate. Let

δ_{11} – the bar separation at R_1 caused by a unit value of R_1.

δ_{22} – the bar separation at R_2 caused by a unit value of R_2.

$\delta_{12} = \delta_{21}$ – the separation at R_1 (or R_2) caused by a unit value of R_2 (or R_1).

δ_{10} – the separation at R_1 due to the load.

δ_{20} – the separation at R_2 due to the load.

The equation

$$\delta_{11}R_1 + \delta_{12}R_2 + \delta_{10} = 0$$

then states that when all the effects are superimposed in the final solution (it is possible to do this since the equations are linear), the bar separation at R_1 must be zero. The effects to be superimposed are the displacement caused by the load, the displacement caused by a unit value of R_1 multiplied by its actual value, and the displacement caused by a unit value of R_2 multiplied by its actual value. The second of Eqs (4.2) similarly requires that when all effects are superimposed, the separation at R_2 must be zero.

The method of virtual work

$$\delta = \sum_{\text{members}} \frac{F_i^v F_i L_i}{A_i E}$$

is used to compute the coefficients in these equations. In so doing, various combinations of the three solutions computed in this example are used. The term δ_{10} which is the discontinuity at R_1 due to the applied load, for example, uses the zero solution and the one solution in the virtual work expression. In this case $F^v \to F_1$ and $F \to F_0$ since the real structure is the zero load case and the virtual structure corresponds to the one load case. It can similarly be argued that the coefficient δ_{ij} combines the ith and the jth load cases in the virtual work expression. A tabular form is helpful when computing these coefficients.

Step 4. Solve for the redundants. Given the coefficients it is necessary to solve two simultaneous equations for the redundants R_1 and R_2.

$$24.42\,R_1 \quad + 5R_2 - 121.6 = 0$$
$$5R_1 + 24.42\,R_2 - 121.6 = 0$$

Subtracting these equations gives

$$19.42\ (R_1 - R_2) = 0 \qquad R_1 - R_2 = 0$$

Adding the equations gives

$$29.42\ (R_1 + R_2) = 243.2 \qquad R_1 + R_2 = 8.27^{\text{K}}$$

or

$$R_1 = R_2 = 4.13^{\text{K}}$$

Step 5. Compute the final solution and plot the results. This has been done in the example.

4.1.3 The General Case

Step 1. Introduce releases to make the structure statically determinate. The basic idea is to reduce a structure to something which is workable. In this 'workable' implies a statically determinate structure for which it is possible to compute both forces and displacements. In order to do so 'releases' are introduced into the structure. As remarked in Chapter 3, a release is a mechanical device which forces a particular stress resultant called a 'redundant' to be zero. When a release is inserted in a structure it creates, by definition, a discontinuity. In the final solution, the redundants are selected so that the value of each discontinuity is zero.

With regard to this text, releases will be introduced on a trial and error basis. When introduced properly, releases can create a statically determinate structure; when introduced improperly the resulting structure will either be statically indeterminate or unstable. The number of releases required to make a structure statically determinate is called the 'degree of statical indeterminacy, k', For more comments concerning the degree of statical indeterminacy the reader should refer to Appendix 6.

Step 2. Analysis. It is necessary to solve the reduced structure for $k + 1$ loading conditions: k cases of unit loads corresponding to individual redundants and one case of the reduced structure under the applied loads.

Step 3. Set up and solve the superposition equations. In general these equations have the form

$$\delta_{11}R_1 + \delta_{12}R_2 + \delta_{13}R_3 + \cdots + \delta_{1k}R_k + \delta_{10} = 0$$
$$\delta_{21}R_1 + \delta_{22}R_2 + \delta_{23}R_3 + \cdots + \delta_{2k}R_k + \delta_{20} = 0$$
$$\vdots \qquad\qquad\qquad\qquad\qquad\qquad \vdots$$
$$\delta_{k1}R_1 + \delta_{k2}R_2 + \delta_{k3}R_3 + \cdots + \delta_{kk}R_k + \delta_{k0} = 0$$

or more simply in matrix form

$$\delta R + \delta_0 = 0$$

where δ_{ij} is the discontinuity at release i due to a unit value of redundant

j, δ_{io} is the discontinuity at release i due to the applied 'loads' or other external effects such as temperature and settlement, and R_i is unknown value of the i redundant.

To compute the coefficient δ_{ij} again requires combining the i and j solutions in the virtual work expression. As remarked above, each of these equations requires the discontinuity at a specific release to be zero in the final solution where all effects are present. For a structure which is statically indeterminate to the kth degree, the force method requires the solution of k simultaneous equations.

Step 4. Combine solutions and plot the results. As before the final solution is the combination of all effects,

$$\text{final solution} = \text{zero solution} + R_1 \cdot \text{one solution}$$
$$+ R_2 \cdot \text{two solution} + \cdots + R_k \cdot k\text{th solution}$$

4.2 THE DISPLACEMENT METHOD

There is a less common method of structural analysis called the displacement method. In this method 'constraints' are added to the structure until it becomes workable or falls within the realm of some known solutions. The displacements associated with the constraints are then selected so that the fictitious forces associated with the constraints are zero in the final solution. This is most easily explained through examples.

4.2.1 A structure with a single degree of freedom
Example 4.4 describes an application of the displacement method to a simple rigid frame. It can be argued in the following manner. In general a rigid frame joint has three degrees of freedom, a horizontal displace-

---

Example 4.3 Frame analysis using the force method.

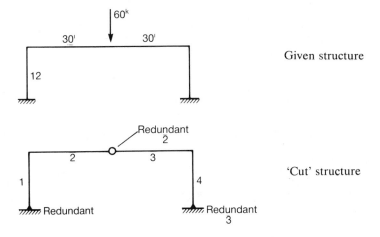

Given structure

'Cut' structure

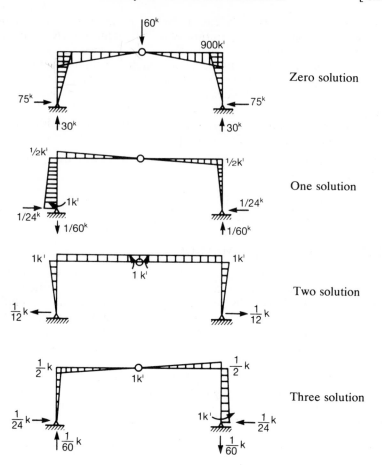

Zero solution

One solution

Two solution

Three solution

Table of Computations

Member	1	2	3	4
L	12'	30'	30'	12'
A	31.8 in²	39.8 in²	39.8 in²	31.8 in²
I	4470 in⁴	7820 in⁴	7820 in⁴	4420 in⁴
F^0	-30^K	-75^K	-75^K	-30^K
F^1	$1/60^K$	$-1/24^K$	$-1/24^K$	$-1/60^K$
F^2	0	$1/12^K$	$1/12^K$	0
F^3	$-1/60^K$	$-1/24^K$	$-1/24^K$	$1/60^K$
F^0F^1L/A	-0.1887	2.355	2.355	0.1887
F^0F^2L/A	0	-4.711	-4.711	0
F^0F^3L/A	0.1887	2.355	2.355	-0.1887
F^1F^1L/A	0.0001048	0.001308	0.001308	0.0001048
F^1F^2L/A	0	-0.002617	-0.002617	0

F^1F^3L/A	−0.0001048	0.001308	0.001308	−0.0001048
F^2F^2L/A	0	0.005234	0.005234	0
F^2F^3L/A	0	−0.002617	−0.002617	0
F^3F^3L/A	0.0001048	0.001308	0.001308	0.0001048
M^0	12′ ▱900	30′ 900▱	30′ ▱900	900▱ 12′
M^1	1▭½	½▱	▱½	½▱
M^2	▱1	1▭	▭1	1▱
M^3	▱½	½▱	▱½	½▭1
$\int \dfrac{M_0 M^1\,dx}{I}$	−0.8054	−0.5754	+0.5754	0.4027
$\int \dfrac{M_0 M^2\,dx}{I}$	−0.8054	−1.726	−1.726	−0.8054
$\int \dfrac{M^0 M^3\,dx}{I}$	0.4027	0.5754	−0.5754	−0.8054
$\int \dfrac{M^1 M^1\,dx}{I}$	0.001566	0.0003197	0.0003197	0.0002237
$\int \dfrac{M^1 M^2\,dx}{I}$	0.0008947	0.0009591	−0.0009591	−0.0004474
$\int \dfrac{M^1 M^3\,dx}{I}$	−0.0004473	−0.0003197	−0.0003197	−0.0004473
$\int \dfrac{M^2 M^2\,dx}{I}$	0.0008949	0.003836	0.003836	0.0008949
$\int \dfrac{M^2 M^3\,dx}{I}$	−0.0004474	−0.0009591	0.0009591	0.0008947
$\int \dfrac{M^3 M^3\,dx}{I}$	0.0002237	0.0003197	0.0003197	0.001566

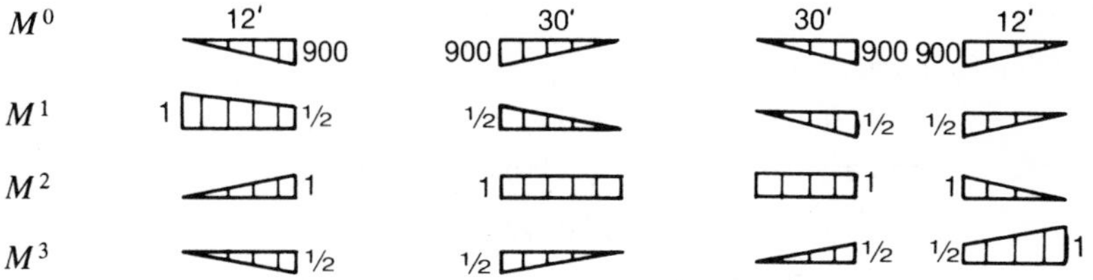

Superposition equations:

$$\delta_{10} + \delta_{11}R_1 + \delta_{12}R_2 + \delta_{13}R_3 = 0$$
$$\delta_{20} + \delta_{21}R_1 + \delta_{22}R_2 + \delta_{23}R_3 = 0$$
$$\delta_{30} + \delta_{31}R_1 + \delta_{32}R_2 + \delta_{33}R_3 = 0$$

Computation of coefficients:

(Note: (1) The factor E cancels and does not appear in the calculations. (2) The units of the coefficients are $K^2\,ft/in^2$.)

$$\delta_{10} = 4.71 - 144 \times 0.4027 = f53.279 = \delta_{30}$$
$$\delta_{20} = -9.422 - 144 \times 5.0628 = -738.46$$
$$\delta_{11} = 0.002826 + 144 \times 0.002429 = 0.3526 = \delta_{33}$$
$$\delta_{22} = 0.010468 + 144 \times 0.009462 = 1.37297$$
$$\delta_{12} = -0.005234 + 144 \times 0.0004474 = 0.0592 = \delta_{23}$$
$$\delta_{13} = 0.002406 - 144 \times 0.001534 = -0.2185$$

(As always symmetry holds, i.e. $\delta_{ij} = \delta_{ji}$.)

Equations in numerical form

$$-53.279 + 0.3526R_1 + 0.0592R_2 - 0.2185R_3$$

$$-738.46 + 0.0592R_1 + 1.37297R_2 + 0.0592R_3$$

$$-53.279 - 0.2185R_1 + 0.0592R_2 + 0.3526R_3$$

solution: $R_1 = R_3 = 169.8$ K$'$, $R_2 = 523.65$ K$'$

Final moment diagram:

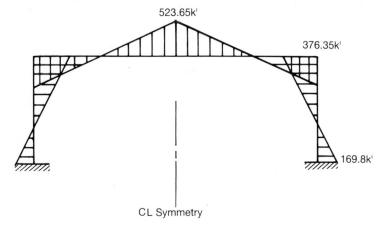

523.65k^l

376.35k^l

169.8k^l

CL Symmetry

Moment at knee $= \dfrac{PL}{4} - 523.65$

$$= \frac{60 \times 60}{4} - 523.65$$

$$= 376.35 \text{ K}'$$

---

ment, a vertical displacement, and a rotation. In this particular case, if member length changes are neglected as it is sometimes common to do, one kinematic degree of freedom remains – the joint rotation θ. When θ is specified, it is possible to compute member forces using only commonly available beam solutions (see Appendix 7).

As a parallel to the force method, the rotation θ can be specified to be zero (i.e. a constraint can be introduced), but to do so requires the application of a fictitious external moment F_{10}. A unit value of θ again requires the application of an external fictitious moment, say F_{11}. In the final solution the value of θ is selected so that this fictitious external moment is zero. Solutions are again superimposed to obtain the final moment diagram.

4.2.2 A structure with two degrees of freedom

Example 4.5 describes a two-degree of freesom structure analysed by the displacement method. In this case a three-span beam is to be solved. This

---

Example 4.4   Rigid frame analysis using the displacement method and neglecting length change.

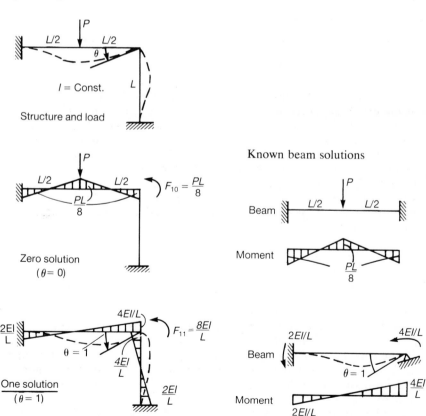

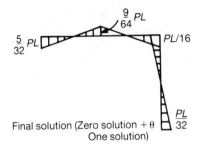

Final solution (Zero solution $+ \theta$
One solution)

Superposition equation:

$$F_{11}\,\delta_1 + F_{10} = 0$$

or

$$\frac{8EI}{L}\theta - \frac{PL}{8} = 0$$

$$\theta = \frac{PL^2}{64EI}$$

is a two-degree of freedom system, since knowing the two beam rotations at the center supports means that the structure has been reduced to well known solutions for single span beams. In order to emphasize the idea of a mechanical constraint to which moments may be applied, fictitious shafts are indicated in the figure (of Example 4.5). As in the case of the force method, a two-degree of freedom requires three analyses, a zero case which provides for the external load and two cases of unit rotations at the constraints. Having these analyses it is then possible to write the two superposition equations which require that the fictitious constraint forces go to zero in the final solution,

$$F_{11}\,\delta_1 + F_{12}\,\delta_2 + F_{10} = 0$$

$$F_{11}\,\delta_1 + F_{22}\,\delta_2 + F_{20} = 0$$

Here δ_1, δ_2 are rotations at constraints 1 and 2, F_{11} is the constraint moment at 1 due to a unit value of δ_1, $F_{12} = F_{21}$ is the constraint moment at 1 (joint 2) due to a unit rotation at joint 2 (joint 1), F_{22} is the constraint moment at joint 2 due to a unit rotation at joint 2, F_{10} is the constraint moment at joint 1 due to load, and F_{20} is the constraint moment at joint 2 due to load.

When the load is applied but the interior joints are not allowed to rotate, a constraint of $wL^2/8$ must be applied externally to the first joint as indicated in the figure (of example 4.5). It follows that

$$F_{10} = -wL^2/8 \qquad F_{20} = 0$$

If a unit rotation is applied to the first contraint, external moments must be applied as indicated in the figure (of Example 4.5). It follows that

$$F_{11} = \frac{7EI}{L} \qquad F_{21} = \frac{2EI}{L}$$

and from symmetry that

$$F_{22} = \frac{7EI}{L} \qquad F_{12} = \frac{2EI}{L}$$

In the case of the example under discussion the superposition equations become

$$\frac{7EI\,\delta_1}{L} + \frac{2EI}{L}\delta_2 - \frac{wL^2}{8} = 0$$

$$\frac{2EI}{L}\delta_1 + \frac{7EI}{L}\delta_2 = 0$$

Adding equations it follows that

$$\frac{9EI}{L}(\delta_1 + \delta_2) = \frac{wL^2}{8}$$

Subtracting equations it follows that

$$\frac{5EI}{L}(\delta_1 - \delta_2) = \frac{wL^2}{8}$$

and that

$$\delta_1 + \delta_2 = \frac{wL^3}{72EI} \qquad \Rightarrow \qquad \delta_1 = \frac{7\,wL^3}{360\,EI}$$

$$\delta_1 - \delta_2 = \frac{wL^3}{40EI} \qquad\qquad \delta_2 = -\frac{2\,wL^3}{360\,EI}$$

The final solution is shown in the figure.

4.2.3 A three-degree of freedom structure and the general case

It is now possible to formulate a general statement of the displacement method:

Step 1. Problem definition. Given a structure it is first necessary to define the number of constraints c required to reduce the structure to a workable form. This number is referred to as the degree of kinematic indeterminacy. It should be pointed out that while the degree of statical indeterminacy can be shown to be invariant for a given structure, the degree of kinematic indeterminacy depends upon the type of analysis to be performed and the types of solutions which are available to work with.

Step 2. Analysis. For a structure which is kinematically indeterminate to the cth degree it is necessary to perform $c + 1$ analyses, one for the loaded structure and c cases in which unit values of the unknown displacements are applied.

Step 3. Set up and solve the c equations of superposition. Let

F_{ij} – fictitious force at the ith constraint due to the effect of a unit value of δ_j with all other constraints set to zero.

F_{io} – the value of the fictitious force at the ith constraint due to load.

δ_i – unknown displacement of the ith constraint.

The superposition equations then appear as

$$F_{11}\,\delta_1 + F_{12}\,\delta_2 + \cdots + F_{1c}\,\delta_c + F_{10} = 0$$
$$F_{21}\,\delta_1 + F_{22}\,\delta_2 + \cdots + F_{2c}\,\delta_c + F_{20} = 0$$
$$\vdots \qquad\qquad\qquad\qquad\qquad \vdots$$
$$F_{c1}\,\delta_1 + F_{c2}\,\delta_2 + \cdots + F_{cc}\,\delta_c + F_{c0} = 0$$

This $c \times c$ system can be written more compactly as

$$\sum_j F_{ij}\,\delta_j + F_{i0} = 0$$

or

$$F\,\delta + F_0 = 0$$

Each of these equations states simply that the fictitious forces associated with each constraint must be zero in the final solution; the final solution is a superposition of the fictitious force due to load when all the constraint displacements are zero plus a term for each unit constraint multiplied by the actual value of that displacement. The coefficients F_{ij} are computed as part of the analysis of Step 2.

Step 4. Construct the final solution. The final solution is obtained by superposition as

Final solution = zero solution + (one solution) · δ_1 + (two solution) · δ_2 + $\cdots$ + (cth solution) · δ_c

Example 4.6 describes the common rigid frame problem of a single frame

Example 4.4

Given problem

E, I, L are constant

Constraints

Zero solution

Load and zero constraints

One solution

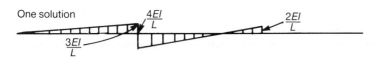

Unit rotation at constraint 1

Two solution

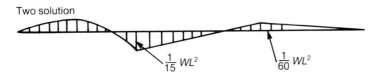

Combining solutions: (moments)

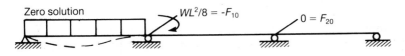

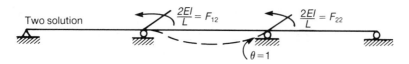

One solution

$\dfrac{2EI}{L} = F_{11}$ $\dfrac{2EI}{L} = F_{21}$

$\theta = 1$

multiply $x \dfrac{7}{360} \dfrac{WL^3}{EI}$

Two solution

$\dfrac{2EI}{L} = F_{12}$ $\dfrac{2EI}{L} = F_{22}$

$\theta = 1$

multiply $x - \dfrac{2}{360} \dfrac{WL^3}{EI}$

bent subjected to a uniform load. Kinematically this structure has six degrees of freedom, two displacements and a rotation at each joint. If the member length changes are neglected, three degrees of freedom remain, two joint rotations and a horizontal (sideway) displacement.

The superposition equations for a three degree of freedom system are

$$F_{11}\,\delta_1 + F_{12}\,\delta_2 + F_{13}\,\delta_3 + F_{10} = 0$$
$$F_{21}\,\delta_1 + F_{22}\,\delta_2 + F_{23}\,\delta_3 + F_{20} = 0$$
$$F_{31}\,\delta_1 + F_{32}\,\delta_2 + F_{33}\,\delta_3 + F_{30} = 0$$

Using the analysis given in the example these equations become

$$\frac{EI}{L} \times \begin{bmatrix} 8 & 2 & \dfrac{6}{L} \\[2mm] 2 & 8 & \dfrac{6}{L} \\[2mm] \dfrac{6}{L} & \dfrac{6}{L} & \dfrac{24}{L^2} \end{bmatrix} \begin{bmatrix} \delta_1 \\[2mm] \delta_2 \\[2mm] \delta_3 \end{bmatrix} = \begin{bmatrix} 1 \\[2mm] -1 \\[2mm] 0 \end{bmatrix} \times \frac{wL^2}{12}$$

Anticipating the problem's symmetry it is convenient to set $\delta_1 = -\,\delta_2$ in order to reduce the work of solution. An equivalent procedure is to rewrite the system matrix as

$$\frac{EI}{L} \times \begin{bmatrix} 10 + \dfrac{12}{L} & & \\ + & 6 & + \\ \dfrac{6}{L} & + & \dfrac{24}{L^2} \end{bmatrix} \begin{bmatrix} \delta_1 + \delta_2 \\ \delta_1 - \delta_2 \\ \delta_3 \end{bmatrix} = \begin{bmatrix} 0 \\ -2 \\ 0 \end{bmatrix} \frac{wL^2}{12}$$

$$\delta_1 + \delta_2 = 0 \qquad \delta_1 - \delta_2 = -\frac{wL^3}{36EI} \qquad \delta_3 = 0$$

and

$$\delta_1 = -\delta_2 = -\frac{wL^3}{72EI}$$

It is now possible to go back and find the moment diagram given the displacements.

Example 4.5 A rigid frame solved by the displacement method neglecting member length change.

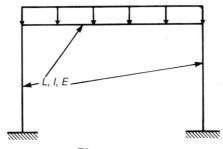

Given structure

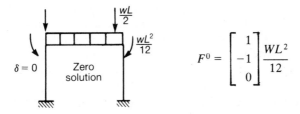

$$F^0 = \begin{bmatrix} 1 \\ -1 \\ 0 \end{bmatrix} \frac{WL^2}{12}$$

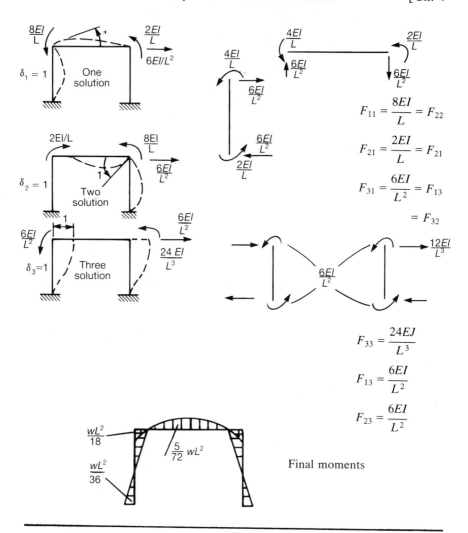

$$F_{11} = \frac{8EI}{L} = F_{22}$$

$$F_{21} = \frac{2EI}{L} = F_{21}$$

$$F_{31} = \frac{6EI}{L^2} = F_{13}$$

$$= F_{32}$$

$$F_{33} = \frac{24EJ}{L^3}$$

$$F_{13} = \frac{6EI}{L^2}$$

$$F_{23} = \frac{6EI}{L^2}$$

Final moments

4.2.4 A pile problem

The pile problem of Example 4.7 shows the power of the displacement method. In this case, bearing piles transfer a load through a soil layer to a bearing stratum below. It is commonly assumed that the piles act like truss bars, that the pile cap is rigid, and that the soil does not exert any frictional force on the piles. As the number of piles increases, this type of structure can become highly statically indeterminate. The interesting point is that this structure has three kinematic degrees of freedom represented by the three degrees of freedom of the pile cap – independent of

Example 4.6   Example of the displacement method – a pile problem.

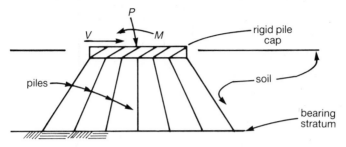

Typical case:

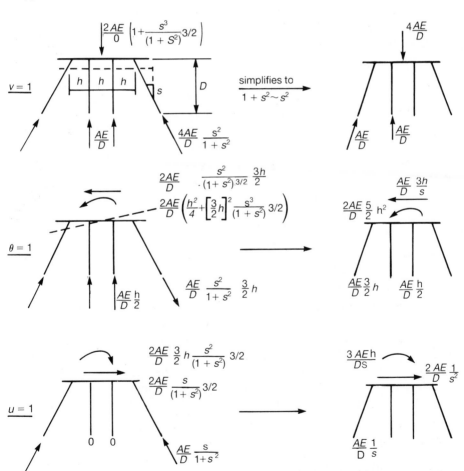

Response to a horizontal load v. Simplified case:

$$\frac{AE}{D} \begin{bmatrix} 4 & & \\ & 5h^2 & -\frac{3h}{s} \\ & -\frac{3h}{s} & \frac{2}{s^2} \end{bmatrix} \begin{bmatrix} v \\ \theta \\ u \end{bmatrix} + \begin{bmatrix} 0 \\ 0 \\ -v \end{bmatrix} = \begin{bmatrix} 0 \\ 0 \\ 0 \end{bmatrix}$$

last two equations uncouple

$$5h^2\theta - \frac{3h}{s}u = 0 \Rightarrow u = \frac{5sh}{3}\theta$$

$$-\frac{3h}{s}\theta + \frac{2}{s^2}u = V \Rightarrow \theta\left(-\frac{3h}{s} + \frac{2}{s^2}\frac{5sh}{3}\right) = V\frac{D}{AE}$$

$$\theta\frac{1}{3}\frac{h}{s} = \frac{VD}{AE} \Rightarrow \theta = \frac{3s}{h}\frac{VD}{AE} \Rightarrow u = \frac{5sh}{3}\frac{3s}{h}\frac{VD}{AE} = 5s^2\frac{VD}{AE}$$

Pile reactions:

Vertical pile $\dfrac{AE}{D}\dfrac{h}{2}\dfrac{3s}{h}\dfrac{VD}{AE} + 0 = \tfrac{3}{2}sV$

Batter pile $\dfrac{AE}{D}\dfrac{3}{2}h\dfrac{3s}{h}\dfrac{VD}{AE} - \dfrac{AE}{D}\dfrac{1}{s}5s^2\dfrac{VD}{AE} = -\tfrac{1}{2}Vs$

Summary:

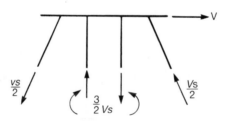

the number of piles used.

The example shows the analysis of a four pile group subjected to three displacements, a vertical displacement, a rotation, and a horizontal displacement and considers a simplified case of batter piles with small slope. An analysis is presented for the simplified problem subjected to a lateral load V. The results are predictable to the extent that each batter pile carries half the load V but causes a resultant moment within the pile group; this moment is balanced by the couple of the two center piles.

4.3 EXERCISES

1 Compute and plot the moment diagram for the frame shown using:

(a) The force method.
(b) The displacement method (neglecting length change)

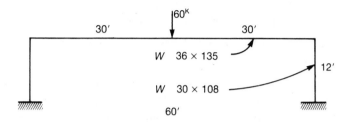

From the *AISC Manual*:

$$W36 \times 135: \quad A = 39.8 \text{ in}^2$$
$$I = 7820 \text{ in}^4$$

$$W30 \times 108: \quad A = 31.8 \text{ in}^2$$
$$I = 4470 \text{ in}^4$$

Note:
(1) Symmetry can be used to simplify calculations.
(2) This problem offers the reader an opportunity to check the common assumption that member length has a negligible effect in the analysis of rigid frames.

2 *Extensions.* Both the force and the displacement method can be extended easily to include such effects temperature, lack of fit, settlement . . . These effects simply appear in the 'forcing terms' F_{i0} and δ_{i0} and do not affect the remainder of the analysis. Put another way, these effects produce discontinuities and constraint forces but they do not affect the manner in which compensation for these discontinuities and constraint forces is made.

For example, suppose that the upper chord members of the truss of Example 4.2 are heated (by the sun) 60°F. This produces a thermal length change in both bars of

$$\Delta L = \alpha . L . \Delta T = 0.65 \times 10^{-5} . 10 . 60$$
$$= 39 \times 10^{-4} \text{ ft}$$

where the coefficient of thermal expansion α is 0.65×10^{-5} per °F for steel. The term $F_{10} = F_{20}$ can now be computed using the virtual work expression

$$\delta = \sum_{\text{members}} F_i^v \Delta L_i = -\frac{1}{\sqrt{2}} . 39 \times 10^{-4} = -27.6 \times 10^{-4}$$

(see Example 4.2). The term 27.6×10^{-4} simply replaces $121.6/E$ in this example. It follows that (see page 000)

$$29.42(R_1 + R_2) = 27.6 \times 10^{-4}E = 27.6 \times 10^{-4} \times 30 \times 10^3$$
$$R_1 + R_2 = 2.81 \text{ K}$$
$$R_1 = R_2 = 1.4 \text{ K}$$

Once the redundants have been computed, the other bars follow directly.

Exercise 2(a) Solve the case in which only one of the above bars is heated. Similar remarks hold for the displacement method. For example, suppose that the left support of Example 4.3 settles by some amount d. Since the settlement problem for a single span beam is readily available,

$$M = \frac{6EId}{L^2}$$

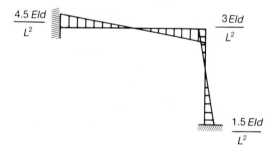

it follows that the fictitious force $F_{10} = -6EId/L^2$ or that

$$\theta = \frac{6EId}{L^2} \div \frac{8EI}{L} = \frac{3}{4}\frac{d}{L}$$

The final moment diagram is therefore

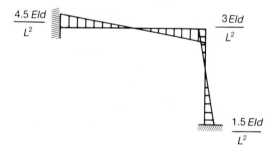

Exercise 2(b) Solve the beam problem of Example 4.4 for the case in which one of the center supports settles an amount d.

3 Analyze the truss shown using the force method.

$$E = 30 \times 10^6 \text{ p.s.i.}$$

All bar areas 2 in^2 except for bar AB whose area is 1 in^2.

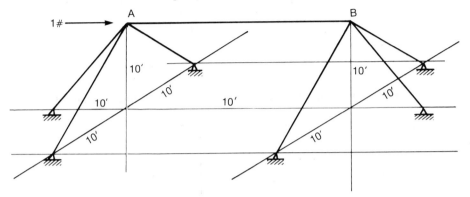

4 Analyze the rigid fram shown using the displacement method. Neglect length change.

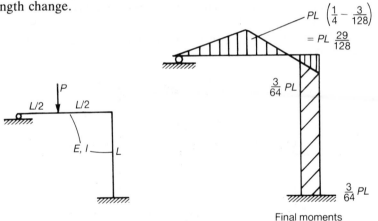

$PL\left(\dfrac{1}{4} - \dfrac{3}{128}\right)$

$= PL\dfrac{29}{128}$

$\dfrac{3}{64} PL$

$\dfrac{3}{64} PL$

Final moments

5 Analyze the structure shown using the force method.

Beam: $A = 50$ in^2, $I = 1500$ in^4 Truss bars: $A = 10$ in^2

Partial answer: Force in bar DB is 4.48^K.

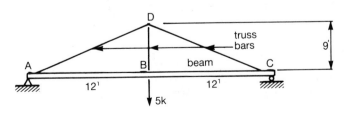

6 Analyze the rigid frame shown using the force method. Partial answer: $R = 17^K$.

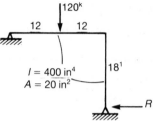

7 Analyze the structure shown using the force method. Assume $A \to \infty$.

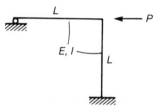

8 Analyze the structure shown using the force method. Partial answer: Force in bar $AE = -7.91^K$.
$E = 29 \times 10^6$ p.s.i. Area of all truss bars: 10 in². Beam: $A = 50$ in², $I = 1500$ in⁴.

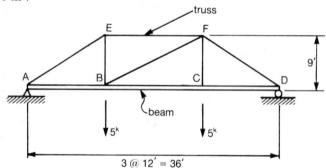

9 Analyze the structure shown using the force method.

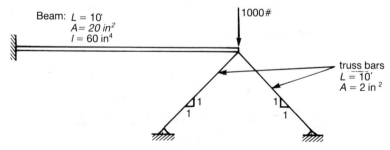

Solution:

$$\delta_{01} = \delta_{02} = \tfrac{1}{72} \times 7.07 \times 10 \times \tfrac{2}{3} \times 10\,000\frac{144}{60E}$$

$$= \frac{565\,600}{E}$$

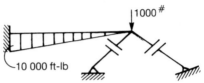

-10 000 ft-lb

$$\delta_{11} = \tfrac{1}{2} \times 7.07 \times 10 \times \tfrac{2}{3} \times 7.07 \times\frac{144}{60E} + \frac{10}{2E}[0.707^2 + 1^2]$$

$$= \delta_{22} = \frac{407.5}{E}$$

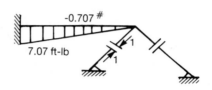

7.07 ft-lb

$$\delta_{12} = \tfrac{1}{2} \times 7.07 \times 10 \times \tfrac{2}{3} \times 7.07 \times\frac{144}{60E} - \frac{10}{2E}(0.707)^2$$

$$= \frac{397.5}{E}$$

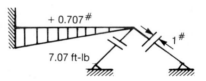

7.07 ft-lb

$$\frac{565\,600}{E} + \frac{407.5}{E}R_1 + \frac{397.5}{E}R_2 = 0$$

$$R_1 = R_2$$

$$\frac{565\,600}{E} + \frac{397.5}{E}R_1 + \frac{407.5}{E}R_2 = 0$$

$$R_1 = -\frac{565\,000}{805} = -702.6\text{ lb}$$

truss reaction = $702.6 \times 2 \times 0.707 = 993.5$

Beam reaction = $1000 - 993.5 = 6.5$ lb

Solution

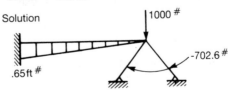

Check the solution using the displacement method.
For a cantilever beam the tip displacement is

$$\delta = \frac{PL^3}{3EI}$$

For a unit vertical displacement each bar shortens by $\dfrac{1}{\sqrt{2}}$

Force required for a unit displacement of the joint

$$\text{Force} = \frac{3EI}{L^3} + \frac{AE}{L} = E\left(\frac{3 \times 60}{1000 \times 144} + \frac{2}{10}\right) = 0.20\ 125E$$

$$\text{Beam shear} = \frac{0.00\ 125}{0.20\ 125}$$

$$\times 1000 = 6.2\ \text{lb} \Rightarrow \text{max beam moment} = 62\ \text{lb-ft}$$

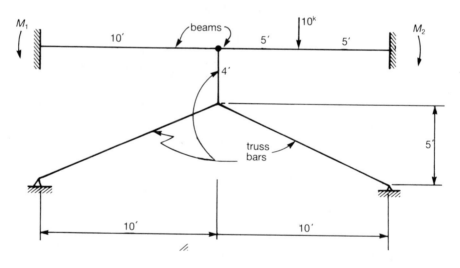

Analyze the structure shown using the force method. Check your solution using the displacement method.

truss bars: $A = 1\ \text{in}^2$ beams: $A = 10\ \text{in}^2$, $I = 1000\ \text{in}^4$

partial answer: $M_1 = 8.92\ \text{K}'$, $M_2 = 27.7\ \text{K}'$

11 Solve the rigid frame shown using the force method.

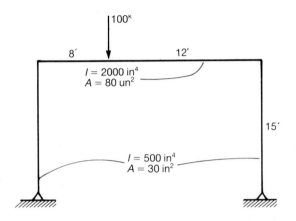

12 Solve the rigid frame shown using the force method. Assume $A \to \infty$.

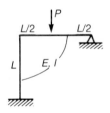

13 Analyze the bracket (rigid frame) shown. All members are 4″ standard steel pipe.

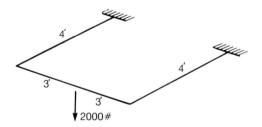

14 Resolve Example 4.3 for the case in which the right support settles by an amount '*d*'.

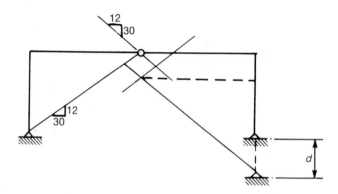

Hint: Given the settlement '*d*', the discontinuities δ_{10}, δ_{20}, and δ_{30} can be computed using simple rigid body mechanics (see below). With these three terms known, the analysis proceeds exactly as given in the example.

$$\delta_{10} = \frac{d}{2 \times 30} \qquad \delta_{30} = -\frac{d}{2 \times 30} \qquad \delta_{20} = 0$$

CHAPTER 5

Plastic analysis

5.1 INTRODUCTION

Thus far this text has been principally concerned with questions of internal forces and reactions generated by external loads. That is, in all cases external loads have been given and the effects of these loads computed. In this chapter a different situation is considered. Here the problem will be to *find* a (collapse) load given a certain description of a structure.

The term *plastic analysis* derives from a simple mode of failure in which *plastic hinges* form within a structure. The typical scenario of plastic collapse has four steps to it (see Figure 5.1):

(1) *Initial elastic response.* Over some limited range of load the behavior of the common structural materials is elastic. The beam equations of Chapter 3 can therfore be used to compute the moments indicated in Step 1 of Fig. 5.1 (Note that in this case the support moments are twice the center moment).

(2) *Plastic hinge formation.* At this point it will be assumed that every beam has a moment 'capacity' μ which cannot be exceeded. It will also be assumed that the beam is elastic up to the point at which the first hinge forms and that hinges form suddenly. (This will be seen later to be only approximately true.) It that is the case then the moment diagram of Step 1 can simply be scaled up to obtain the moment diagram of Step 2. It should also be noted here that while the moment capacity has been reached at 2 points in the beam, the beam can still carry additional load.

(3) *Moment redistribution.* If the load is to increase without an increase in the support moments (which have reached their capacity) it follows that the center moment must increase and that the ratio of the support and center moments must change. That is, the centre moment can no longer remain at half the support moments.

(4) *Plastic collapse.* As the load continues to increase, the center moment eventually reaches its capacity μ. At this time the beam is said to be

Step 1   Initial elastic response

Loaded beam

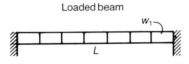

moment diagram

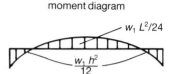

Step 2   Formation of plastic hinges

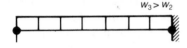

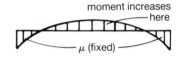

Step 3   Moment of redistribution

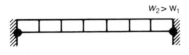

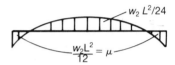

Step 4   Plastic collapse

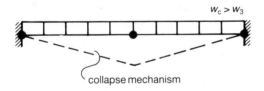

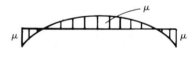

Fig. 5.1 – The scenario of plastic collapse.

at the point of collapse since any increase in load would then require the moment capacity to be exceeded somewhere. Note also that for small deformations a beam with three hinges is a mechanism. (Statics, of course, requires that $2\mu = \omega_c L^2/8$ or that the *collapse load* $\omega_c = 16\mu/L^2$.)

This scenario will be argued in detail below.

It will be seen in the next section that the phenomenon of plastic collapse makes some demands upon material properties. For example, if you compare a paper clip to a piece of chalk, you can see that it is quite

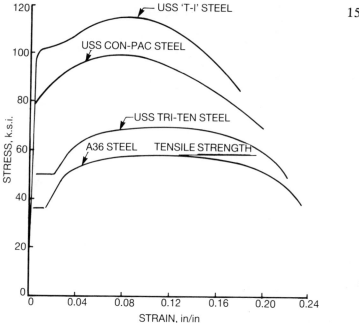

Typical stress-strain curves for structural steels having specified minimum tensile properties.

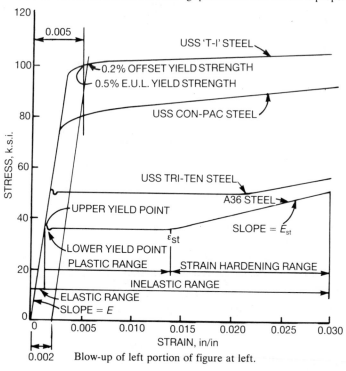

Blow-up of left portion of figure at left.

STEEL

(Taken from *USS Steel Design Manual* by R. L. Brockenbrough and B. G. Johnston, United States Steel Corp, 1968)

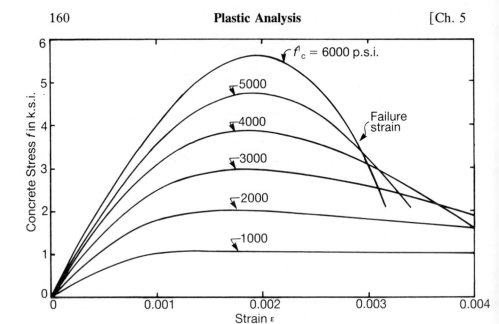

CONCRETE

(Taken from 'Ultimate Strength of Over-Reinforced Beams', by Ladislav B. Kriz and Seng-Lip Lee, *Proc. ASCE*, 86, EM3, June 1960, p. 98.)

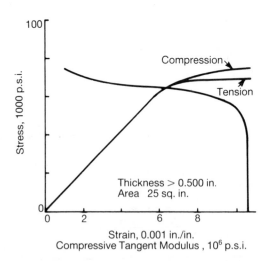

ALUMINUM

(Taken from MIL-HOBKS, August 1962, 'Metallic Materials and Elements for Flight Vehicle Structures', US DoD.)

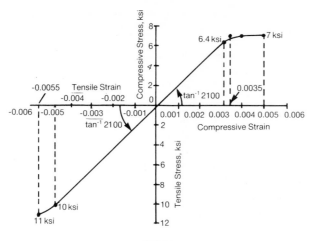

WOOD

Stress strain diagram in bending of clear wood of a certain specific gravity at a given moisture content and temperature (Taken from *Wood Engineering* by German Gurfinkel, published by Southern Forest Products Assoc., New Orleans.)

Fig. 5.2 – Typical stress–strain curves.

easy to make a 'kink', the sign of a plastic hinge, in the wire of the paper clip but that the chalk is brittle and simply breaks. In this vein, it will be shown below that plastic analysis requires a ductile material. However, as Fig. 5.2 indicates, the ductility of the common structural materials varies considerably:

Steel. The designer has many options available to him when working with structural steel. While the so-called A36 steel is by far the most common 'structural steel', there are situations in which he may wish to use 'high strength' steels to reduce weight or cost.

Concrete. As indicated in the figure many strengths of concrete are available to the structural engineer. In the past concrete with a strength of 3000 psi has been common but present day economies tend to make higher strengths attractive. (Note that the stress-strain curve shown is one of compression. The strength properties of concrete in tension are small and frequently neglected.)

Aluminium. While aluminium has been used in buildings and bridges, it is not common to do so. As a rule of thumb, aluminium is about one-third that of steel at comparible strengths. Its Young's modulus is about one-third that of steel (which implies some possible buckling problems) and it is more difficult than steel to join together at connections. But its lack of use is for the most part a matter of economics.

Wood. Wood is, of course, the work-horse material for light construction but somewhat out of vogue these days for heavy construction. However, depending upon the economics of the moment it is common to interchange (laminated) wood and steel framing.

In most simple terms steel is a highly ductile material for which plastic analysis is most commonly used. Concrete (shown in compression in Fig. 5.2) is a relatively brittle material but the proper use of reinforcement can make reinforced concrete a more ductile material. Plastic analysis is not commonly used for aluminium and wood.

The remainder of this chapter is devoted to examining the phenomenon of plastic collapse more carefully and to developing some analysis procedures. Certainly it is appropriate to examine the formation of plastic hinges more carefully. Beyond this the scenario outlined in Fig. 5.1 is quite general but as a rule not easy to follow in a complex structure. For that reason some space will be devoted to developing formal theorems and two methods of plastic analysis.

5.2 THE CONCEPT OF A PLASTIC HINGE

Having discussed briefly some of the realities of materials, it is now time to move on to the common idealizations of plastic analysis. Steel is of course the prototype material of plastic analysis and as such deserves one final look. First, it has a failure strain 25% (see Fig. 5.2) which is quite extraordinary. Fig. 5.3 indicates other properties. Within the platic range

Ideal elastic plastic material

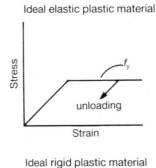

Ideal rigid plastic material

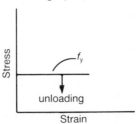

Fig. 5.3 – Idealized stress–strain curves.

mild steel (A36) is really very close to an ideal elastic-plastic material. Futhermore, it is common to neglect the strain hardening which develops a tensile strength considerably above the yield point. The argument is simply that the strains encountered in practice do not carry steel into the strain hardening region. As a result, the analysis which follows will use either an elastic-plastic material model or a rigid-plastic model.

Fig. 5.4 describes the formation of a plastic hinge as the moment increases at any point in a beam of an ideal elastic-plastic material using the assumption that plane sections remain plane. For a small moment M, the

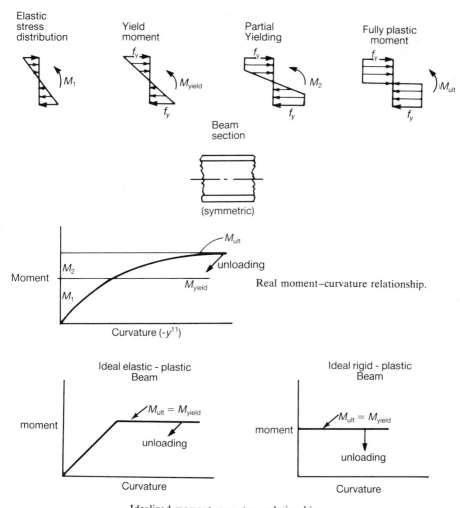

Fig. 5.4 – Formation of a plastic hinge.

entire cross-section remains elastic and the stresses can be computed directly using beam theory; for some larger moment M_{yield} yielding first occurs in the extreme fibers; at some still larger moment M_2 yielding proceeds partially across the section; and M_{ult} is an asymptotic state at which there is yielding across the entire section. The actual moment–curvature relationship is also indicated in this figure.

It is usual at this point to make the assumption of first an ideal elastic-plastic moment–curvature relationship and finally the assumption of a rigid-plastic moment–curvature relationship. The former assumption is based upon the fact that the ratio of M_{yield}/M_{ult} is section dependent but commonly is quite close to 1 for wide flange sections where the material is concentrated in the flanges. Consider two cases:

Rectangular section:

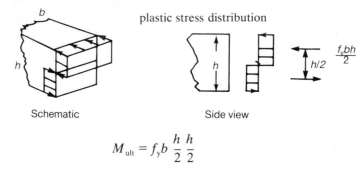

Schematic   Side view

$$M_{ult} = f_y b \frac{h}{2} \frac{h}{2}$$

At first yielding, the elastic stress in the extreme fiber is f_y

$$\sigma = \frac{M_c}{I} = \frac{\sigma M}{bh^2} = f_y \Rightarrow M_{yield} = \frac{f_y bh^2}{6}, \qquad M_{yield}/M_{ult} = 2/3$$

Wide flange section:

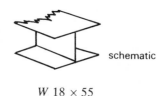

schematic

$W\ 18 \times 55$

Properties for this section are

$$S_x = 98.4\ in^3 \qquad Z_x = 112\ in^3 \quad (see\ AISC\ Manual)$$

$$M_{yield} = f_y S_x = f_y\ 98.4 \qquad M_{ult} = f_y Z_x = f_y\ 112$$

$$M_{yield}/M_{ult} = 98.4/112 = 0.878 \quad see\ also\ Exercise\ \mathbf{1},\ page\ 182.$$

Neglecting the elastic portion of the ideal elastic-plastic moment–curvature relationship and using the rigid-plastic moment–curvature relationship is based upon the argument that during most of the response to a real design situation the elastic response is insignificant. That must, of course, be verified in specific cases.

In any case, when M_{ult} has been reached at any point of a beam, additional rotation cannot increase the moment. A situation where rotation is not accompanied by a change in moment is of course the classic case of a hinge – in this case a *plastic hinge*. It should be noted parenthetically that the beam behaves elastically with respect to *decreasing* curvature.

5.3 PROPORTIONAL LOADING

For a structure which is subjected to more than one load, the question arises concerning the manner in which the loads vary as the failure load is approached. This is, of course, simply a matter of clearly defining the loading of the structure.

One common way to deal with this problem is through proportional loading (Fig. 5.5). With proportional loading the *ratios* of the various loads

Arbitrary loading   Proportional loading

remain fixed. This allows the loading on any structure to be described by a single parameter. Formally then any load vector $\mathbf{P}_i$ can be described as some fixed value $\mathbf{P}_i^0$ times some scalar λ which is allowed to vary, i.e.

$$\mathbf{P}_i = \lambda \mathbf{P}_i^0$$

Proportional loading is assumed throughout this chapter.

5.4 MOMENT REDISTRIBUTION

One of the interesting and useful aspects of plastic analysis is the *moment redistribution* which occurs in structures after they have begun to yield. It is characteristic of elastic structures that large moments tend to develop at places like supports. This frequently results in a poor utilization of materials since most of the structure may then be under-stressed. Fig. 5.1 indicates a case in point. The elastic solution for a fixed-ended uniform beam subjected to a uniformly distributed load w_1 has support moments of twice the magnitude of the center span moment. In terms of design this is a wasteful situation. At *plastic* collapse, on the other hand, the support and center moments are in fact equal for a uniform beam.

Moment redistribution has implications well beyond this simple situation. For example, joints in rigid frames play the same role as supports in beams and similarly tend to 'collect' large moments under elastic behavior. Using plastic analysis, sections of equal strength can carry equal moments at failure. This fact has considerable implications for problems of optimal design.

5.5 ANALYSIS THEOREMS

In most cases following the loads step by step to the point of collapse is too complex a process to be generally useful as an analysis tool. In order to develop a somewhat more sophisticated approach to plastic analysis, two classical theorems – an upper bound theorem and a lower bond theorem – are discussed in this section. These theorems will subsequently form the basis of systematic and workable methods of plastic analysis.

5.5.1 The Virtual Work Equation
The upper and lower bound theorems make use of a so-called work energy equation which in turn can be derived using the equations of virtual work for a rigid body. It is shown in Chapter 1 that the virtual work of a body in equilibrium under rigid body motion is zero. In order to derive the work energy equation it is first necessary to extend this virtual work equation to the case of a system of several rigid bodies which are joined by pins. In this case the virtual work equation generalizes to

$$\sum_{\text{loads}} \mathbf{P}_i \cdot \boldsymbol{\delta}_i + \sum_{\text{torques}} \mathbf{T}_i \cdot \boldsymbol{\omega}_i + \sum_{\text{hinges}} M_i \theta_i = 0 \qquad (5.1)$$

Here M_i is the moment associated with the ith hinge and θ_i is the angle change (discontinuity) associated with the ith hinge.

Eq. (5.1) can be obtained by adding the cirtual work equations of the several rigid bodies which are joined by the pins. The *forces* on the pins do not appear in the final equation since the work they do cancels out in the sum. The moments at the pins may, however, do work since each of the elements may not rotate the same amount. These angular discontinuities give rise to the last term in Eq. (5.1).

Fig. 5.6 shows a simple rigid frame which is used here for two reasons. There is first an equilibrium equation

$$PL + (M_1 + M_2 + M_3 + M_4) = 0$$

between the column moments and the lateral load which can be obtained directly using Eq. (5.1). This figure also shows how this equation can

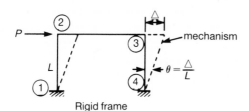

Rigid frame

Free body diagrams

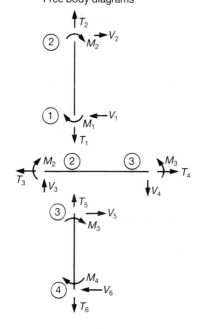

Virtual work equations

$$M_1\theta + M_2\theta + V_2\Delta = 0$$

$$T_4\Delta - T_3\Delta = 0$$

$$M_3\theta + M_4\theta + V_5\Delta = 0$$

Adding virtual work equations

$$(M_1 + M_2 + M_3 + M_4)\theta + (V_2 + V_5 + T_4 - T_3)\Delta = 0$$

$$P = V_2 - T_3$$

$$T_4 + V_5 = 0$$

Equilibrium combined with virtual work $(\theta = \Delta/L)$

$$(M_1 + M_2 + M_3 + M_4)\frac{1}{L} + P = 0$$

Fig. 5.6 – Example of the use of the virtual work equation for rigid bodies joined by pins.

be obtained directly from the virtual work equations of the pieces of the frame and in this sense illustrates the derivation of Eq. (5.1).

It should be emphasized that virtual work equations in general only reflect conditions of statics (equilibrium). The clever use of virtual work frequently provides a simple means of dealing with some situations of complex geometry as in the case of Fig. 5.6 where the 'sidesway' equation can be produced directly.

5.5.2 The Work Energy Equation
The work energy equation states that:

> At plastic collapse, the work done by the external loads equals the energy absorbed at the hinges for a small virtual displacement using the real collapse mechanism.

This equation is simply a construct which can be obtained from the virtual work equation for rigid bodies joined by pins (Eq. (5.1)). That equation was relatively easy to write down because no effort was made to carefully define what the discontinuity θ_i at pin i was precisely to be. The work energy equation starts with Eq. (5.1) and assumes that *at plastic collapse* the moment always opposes the rotation (i.e. dissipates energy). The term $M_i \theta_i$ then can be written as $-\mu_i |\theta_i|$ where μ_i is the moment capacity at point i.

The virtual work equation (Eq. (5.1)) then becomes the work energy equation

$$\sum_{\text{loads}} \mathbf{P}_i \cdot \boldsymbol{\delta}_i + \sum_{\text{torques}} \mathbf{T}_i \cdot \boldsymbol{\omega}_i = \sum_{\text{hinges}} \mu_i |\theta_i| \qquad (5.2)$$

Considerable use will be made of this equation in the work which follows. For the time being it will simply be noted that it can be applied to the beam of Fig. 5.7 and used to compute the collapse load.

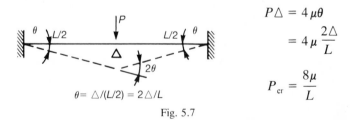

$$P\triangle = 4\mu\theta$$
$$= 4\mu \frac{2\triangle}{L}$$

$$\theta = \triangle/(L/2) = 2\triangle/L$$

$$P_{\text{cr}} = \frac{8\mu}{L}$$

Fig. 5.7

In what follows, the two analysis theorems will be discussed. In both discussions a distinction will be made between *trial* moment diagrams and collapse mechanisms and *real* ones. In terms of analysis, the difference comes through the fact that in a *real* collapse mechanism the moment capacity is reached at a plastic hinge (by definition) and that the moment opposes the rotation. This may not be the case for a trial mechanism. In any case the reader should watch carefully for this distinction which is the key to the following discussions.

5.5.3 The Lower Bound Theorem

In discussing the following theorems it is convenient to drop the applied torque terms from Eq. (5.2). They are in fact relatively uncommon in practice. In any case the lower bound theorem states that

A load for which it is possible to find a safe and statically admissible moment diagram is less than or equal to the collapse load.

Proof: Using proportional loading each load is described as a scalar λ multiplying a given load ratio $\mathbf{P}_i^0$ as

$$\mathbf{P}_i = \lambda \mathbf{P}_i^0$$

The symbol λ^c will be used to describe the collapse load and $\mathbf{\delta}^c$ the collapse mechanism.

The theorem requires a moment diagram which satisfies the equations of statics and at no point exceeds the capacity of the frame. The virtual work equation (Eq. (5.1)) for this feasible moment diagram and the collapse mechanism is

$$\Sigma\, \mathbf{P}_i \cdot \mathbf{\delta}_i^c + \Sigma\, M_i \theta_i^c = 0$$

or

$$\lambda(\Sigma\, \mathbf{P}_i^0 \cdot \mathbf{\delta}_i^c) = -\Sigma\, M_i \theta_i^c \leqslant \Sigma\, |M_i|\,|\theta_i{}^c|$$

$$\leqslant \Sigma\, \mu_i\,|\theta_i^c| = \lambda^c(\Sigma\, \mathbf{P}_i^0 \cdot \mathbf{\delta}_i^c)$$

In the last step use is made here of the fact that the collapse mechanism and collapse load satisfy the work energy equation. It follows that

$$\lambda \leqslant \lambda^c$$

which completes the proof.

Fig. 5.8 shows an example of a uniform frame for which it is desired to find a lower bound on the collapse load P_{cr} using the lower bound theorem. This can be done by constructing a moment diagram which is safe and satisfies the requirements of statics. Anticipating a hinge at the right knee, the right reaction is assumed to be μ/L and the moment diagram follows directly. The moment diagram will be safe if $PL < 2\mu$ which corresponds to a hinge at the left knee. In view of the lower bound theorem

$$P_{cr} \geqslant 2\mu/L$$

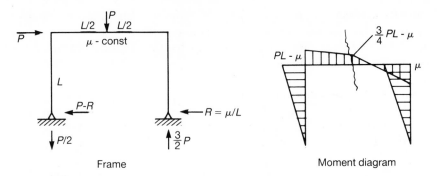

Frame   Moment diagram

Note:
(1) The vertical reactions can be computed directly by taking moments about the supports.
(2) The structure is statically indeterminate to the first degree. R plays the role of a redundant.
(3) It is necessary to compute the moments at the vertical load to see whether or not it is critical.

Fig. 5.8 – An example using the lower bound theorem.

5.5.4 The Upper Bound Theorem

The collapse load can be bounded using the upper bound theorem which states:

A load which satisfies the work energy equation for a given mechanism is greater than or equal to the collapse load.

Proof: Let δ_i and θ_i refer to a given (trial) collapse mechanism. The virtual work equation of the collapse loads on the given mechanism is

$$\lambda^c \sum \mathbf{P}_i^0 \cdot \delta_i = - \sum M_i^c \theta_i$$
$$\leq \sum \mu_i |\theta_i| \quad \text{(since the moment diagram associated with collapse is feasible)}$$
$$= \lambda \sum \mathbf{P}_i^0 \cdot \delta_i \quad \text{(using the work energy equation)}$$

or

$$\lambda^c \leq \lambda$$

Fig. 5.9 indicates a continuation of the problem just discussed showing all possible collapse mechanisms. The implication of the upper bound theorem is that the smallest computed load is the collapse load. For the beam mechanism the work energy equation is

$$P\Delta = 4\mu\theta = 4\mu \frac{\Delta x 2}{L} \qquad \text{or } P^{cr} \leq 8\mu/L$$

For the sidesway mechanism

$$P\Delta = 2\mu\theta = 2\mu\frac{\Delta}{L} \quad \text{or } P^{cr} \leqslant 2\mu/L$$

and for the combined mechanism

$$P\Delta + \frac{P\Delta}{2} = 2\mu\frac{\Delta}{L} + \frac{2\mu\Delta}{3L} \quad \text{or } P^{cr} \leqslant \frac{8}{3}\mu/L$$

Since all possible mechanisms have been tried it can be claimed that the collapse load is $P^{cr} = 2\mu/L$. This is verified by the earlier discussion which indicated $P^{cr} \geqslant 2\mu/L$.

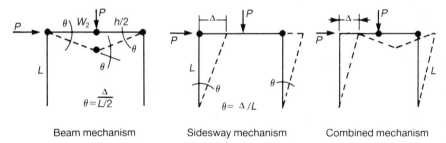

Beam mechanism   Sidesway mechanism   Combined mechanism

Fig. 5.9 – Frame showing all collapse mechanisms

5.6 COMPUTING PLASTIC COLLAPSE LOADS

On the basis of the preceding discussion it is possible to construct several methods for systematic plastic analysis of structures:

(1) *Using the upper bound theorem.* For problems of modest size where it is possible to enumerate all possible collapse mechanisms, the work energy equation may be used. That is, the work energy equation gives an upper bound on the collapse. If all possible mechanisms are tried then the smallest computed collapse load is the actual collapse load.

(2) *Using the lower bound theorem.* It is in some cases possible to follow the load from the initial elastic response to collapse satisfying equilibrium. For a highly redundant structure this can be difficult and does not constitute a workable analysis procedure in most cases.

(3) *Using linear programming.* It is possible (using the lower bound theorem) to set up plastic analysis as a problem in which it is desired to maximize the load λ subject to (a) equilibrium and (b) safety (the moment must be less than the capacity at each point in the structure).
    This is the classical linear programming formulation. It is of

interest since effective algorithms are now available to deal with large-scale linear programming problems.

Examples of these methods are given below (see Examples 5.1, 5.2 and 5.3).

Example 5.1 Find the collapse load P using the upper bound theorem:

(1) List all possible collapse mechanisms.
(2) Write the work energy equation for each mechanism.
(3) Select the smallest load.

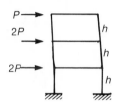

moment capacity

columns – μ
beams – 3μ

Mechanisms:

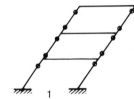 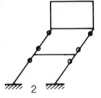 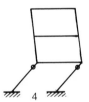

Because of the relative strengths of the beams and columns.

this mechanism governs   not this mechanism

Mechanism 1

$$2P \cdot \Delta + 2P \cdot 2\Delta + P \cdot 3\Delta = \mu12 \times \Delta/h \qquad P = \tfrac{4}{3}\mu/h$$

Mechanism 2

$$2P \cdot \Delta + 2P \cdot 2\Delta + P \cdot 2\Delta = \mu8 \cdot \Delta/h \qquad P = \mu/h$$

Mechanism 3

$$2P \cdot \Delta + P \cdot \Delta = \mu4 \cdot \Delta/h \qquad\qquad P = \tfrac{4}{3}\mu/h$$

Mechanism 4

$$2P \cdot \Delta + 2P \cdot \Delta + P \cdot \Delta = \mu 4 \, \Delta/h$$

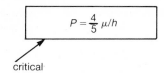

$$P = \frac{4}{5} \mu/h$$

critical

Example 5.2   A case of distributed load.

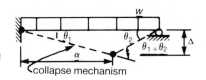

collapse mechanism

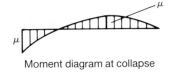

Moment diagram at collapse

Kinematic approach:

Work energy equation: $\frac{1}{2}wL \, \Delta = \mu\theta_1 + \mu(\theta_1 + \theta_2)$

$$\theta_1 = \frac{\Delta}{\alpha} \qquad \theta_2 = \frac{\Delta}{(L - \alpha)} \Rightarrow \frac{1}{2}wL \, \Delta = \mu \left[\frac{2\Delta}{\alpha} + \frac{\Delta}{L - \alpha} \right]$$

$$w = \frac{2\xi}{L} \left(\frac{2}{\alpha} + \frac{1}{L - \alpha} \right)$$

$$\frac{dw}{d\alpha} = 0 \Rightarrow -\frac{2}{\alpha^2} + \frac{1}{(L - \alpha)^2} = 0 \qquad \alpha = 0.586L = \frac{\sqrt{2}L}{1 + \sqrt{2}}$$

$$w = \frac{2\mu}{L} \left(\frac{2}{0.586L} + \frac{1}{0.414L} \right) = \underline{11.66\mu)L^2}$$

Static approach:

$$M'' = -w \Rightarrow M = c_1 + c_2 x - wx^2/2$$

$$M = -\mu \, @ \, x = 0 \Rightarrow c_1 = -\mu$$

find max M: $\dfrac{dM}{dx} = 0 \Rightarrow c_2 - wx = 0 \qquad \bar{x} = c_2/w$

$$M = 0 \, @ \, x = L \qquad 0 = -\mu + c_2 L - \frac{wL^2}{2}$$

2 equations for w, c_2

$$M = \mu \, @ \, x = \bar{x} \qquad \mu = -\mu + c_2 \frac{c_2}{w} - \frac{w}{2} \left(\frac{c_2}{w} \right)^2$$

$$4\mu = \frac{c_2^2}{w} \qquad c_2 = \frac{\mu}{L} + \frac{wL}{2} \Rightarrow 4\mu w = \left(\frac{\mu}{L} + \frac{wL}{2} \right)^2$$

$$w = \underline{11.66\mu/L^2} \quad \text{or} \quad 0.343\mu/L^2 \text{use max. value}$$

Note: The difficulty arises here because the location of the hinge in the beam is not known *a priori*.

Example 5.3 'Linear programming' approach.
Maximize P subject to the conditions that moments at hinges must be less than capacity

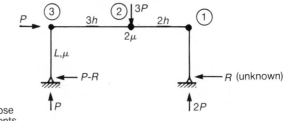

Collapse moments

Check three possible hinges:

(1) $|RL| \le \mu$ or $|R| \le \mu/L$

(2) $|2P \cdot 2L - R \cdot L| \le 2\mu$ or $|4P - R| \le 2\mu/L$

(3) $|(P - R)L| \le \mu$ or $|P - R| \le \mu/L$

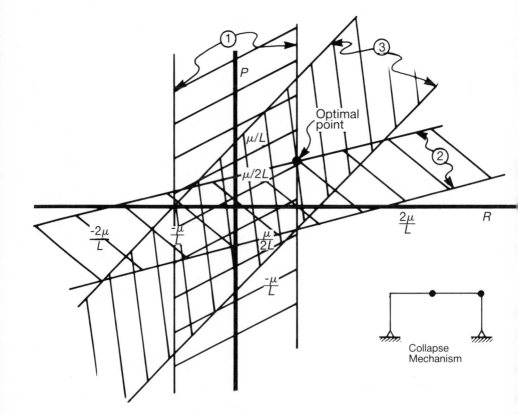

Collapse Mechanism

5.7 EXTENSIONS

5.7.1 Interaction Formulas

As described thus far, plastic analysis has only been concerned with bending. That is, it has been argued that failure will occur when enough plastic hinges have formed and that plastic hinges formed at any point when the moment capacity was reached.

In the more general case, there is an *interaction* between stress resultants at a point. For example, it will be shown below how the presence of axial load reduces the moment capacity of a beam. In general there is an interaction between the six stress resultants at any point of a space beam. A generalized failure criterion describing a 'generalized hinge' would then be an expression involving all six stress resultants.

No attempt will be made here to present a plastic analysis of a space beam but two examples will be used to discuss the effect of axial load on the formation of plastic hinges.

A rectangular beam.   Consider a beam of rectangular cross-section with width b and depth d. The axial load capacity and the moment capacity are simply

$$P_{\text{ult}} = f_y bd \quad \text{and} \quad M_{\text{ult}} = f_y bd^2/4$$

when each is applied separately. When P and M occur in combination they produce the stress distribution indicated in Fig. 5.10 which can be

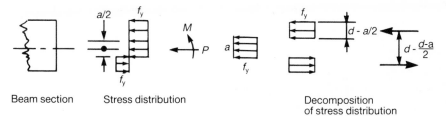

Beam section   Stress distribution   Decomposition
                                                            of stress distribution

Fig. 5.10 – Rectangular Beam

decomposed as shown there. In this case

$$P = f_y ab \quad \text{and} \quad M = f_y b \left(\frac{d - a}{2} \right) \left(d - \frac{d - a}{2} \right)$$

$$= f_y b (d^2 - a^2)/4$$

Eliminating 'a' from these two equations gives the interaction formula

$$\frac{M}{M_{\text{ult}}} + \left(\frac{P}{P_{\text{ult}}} \right)^2 = 1$$

which describes combinations of M and P which correspond to failure. This interaction equation is plotted in Fig. 5.11. Clearly any point under the curve is feasible while points outside the curve violate the assumption of a material yield stress f_y.

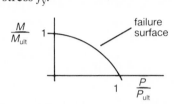

Fig. 5.11 – Interaction Diagram for Rectangular Beam.

A 'flange' beam.   In an attempt to approximate a wide flange beam in which much of the cross-section material is concentrated in the beam flanges, a highly idealized concept of a flange beam is used. This beam has no web and simply consists of two flanges of area $A/2$ spaced at a distance h. The analysis of the preceding section can be duplicated for this beam where

$$P_{ult} = Af_y \quad \text{and} \quad M_{ult} = hf_yA/2$$

The combination of axial load and bending moment is indicated schematically in Fig. 5.12. The critical combination then gives

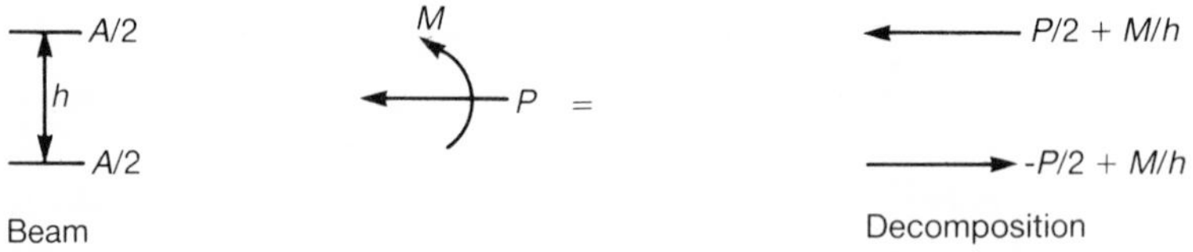

Beam M $P/2 + M/h$

h P =

                                                                        $-P/2 + M/h$

Beam   Decomposition

Fig. 5.12 – Interaction Diagram for a Sandwich Beam.

$$P/2 = \frac{M}{h} = \frac{A}{2}f_y$$

from which the interaction equation follows directly as

$$\frac{P}{P_{ult}} + \frac{M}{M_{ult}} = 1$$

This result is indicated in Fig. 5.13.

Interaction formulas occur often in structural engineering. In general terms, if a design has been prepared for one kind of loading and another kind of loading is to be added, the two loads together are more stringent than either taken alone. Interaction formulas describe just how much more stringent they are. The student will encounter applications of interaction formulas immediately in areas such as basic steel and concrete design when column and beam design procedures are combined to accommodate bending and axial load.

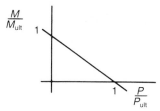

Fig. 5.13 – Interaction Diagram for a Sandwich Beam.

5.7.2 Repeated Loading

For structures such as crane girders and machinery supports where repeated loading is common, consideration must be given to the effect of fatigue. As indicated in Fig. 5.14, under repeated loading materials fail at stress levels well below their tensile strength. In this figure it is worthwhile noting:

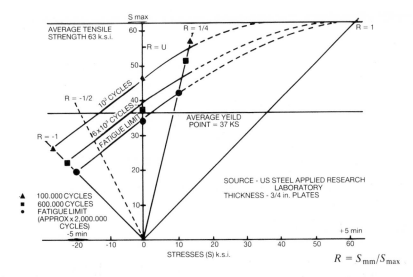

Fig. 5.14 – Fatigue strength of as-received A36 structural carbon steel. (Taken from *USS Steel Design Manual*, R. L. Brockenbrough and B. G. Johnston, United States Steel Corp., 1968.)

(1) In the most simple fatigue test in which the load cycles between 0 and S_{max} (shown along the vertical axis in this figure), the fatigue limit is below the yield point. This means that in order to tolerate a large number of load cycles between 0 and S_{max} it is necessary to keep the stress level below yielding.

(2) As S_{min} approaches S_{max} ($R = 1$) there is really no repeated loading taking place and the fatigue curves asymtotically approach the tensile strength.

(3) The condition of complete reversal of stress ($R = -1$) into the plastic range is thought to be dangerous and something to be avoided when possible. Fig. 5.14 implies that this is the worst possible situation for fatigue and does not, in fact, show the number of cycles to failure for this case.

5.7.3 Shakedown

In addition to concern over fatigue under repeated loading, the engineer must avoid incremental collapse when repeated loading produces stresses above the yield point. Incremental collapse and shakedown are linked in plastic analysis in the following manner. Under repeated loading it can happen that after a few cycles of inelastic response the structure slips into a mode of elastic response for the remaining load cycles. When this happens the structure is said to 'shake-down'. It can also happen that each load cycle produces a small increment of inelastic deformation and that these increments add up to incremental failure. The cases shown in Fig. 5.15 describe some of these possibilities:

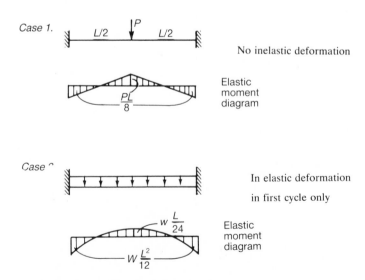

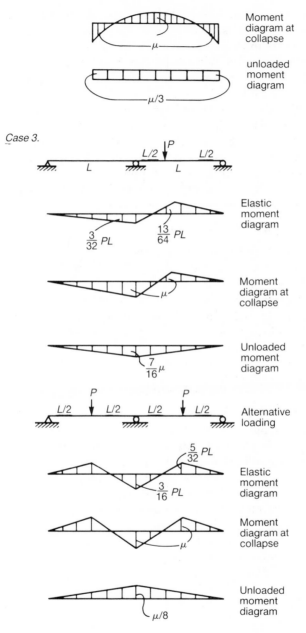

Inelastic deformation every cycle and incremental collapse

Note: Elastic moment diagrams are taken from the *AISC Manual*.

Fig. 5.15 – Load cycling of an elastic-plastic structure.

Case 1. This is an almost trivial case. Since the elastic solution has the same moment at center span as it has at the supports, treated as an elastic plastic beam, the three plastic hinges required for collapse form simultaneously and there is no moment redistribution and no inelastic deformation. Under loading and unloading to the point of collapse, the response of this beam is elastic.

Case 2. This case was discussed earlier in Fig. 5.1 except that unloading was not considered at that time. As before, as the load w is applied the initial response is elastic up to the point at which plastic hinges form at the supports. After these hinges have formed, the response of the structure to additional load is that of a simply supported beam. When the moment capacity is reached at the center of the beam no additional load can be added. The collapse load can be computed as

$$2\mu = wL^2/8 \Rightarrow w = 16\mu/L^2$$

Now there is an important point to be made. The earlier discussions imply that the response of a structure to load reversal is elastic up to the point that hinges form again. Now this fact must be applied carefully. The moment diagram after load reversal is determined by combining the collapse moment diagram with an elastic moment diagram which corresponds to load removal. The support moments after unloading are then

$$\mu = wL^2/24 = \mu - \frac{L^2}{24}\frac{16\mu}{L^2} = \mu/3$$

and the unloaded or residual moment diagram is constant at $\mu/3$ as indicated in the figure. If the load w is again applied and removed the structure behaves elastically up to collapse and there is no additional inelastic deformation. This structure is said to have shaken down.

Case 3. This example is somewhat more complex and thus more difficult to describe. There are first of all two sets of loads as indicated in the Fig. 5.15. Furthermore, it is important to think of four separate steps: a load is put on and it must be taken off; then the other must be put on and it must be taken off. When the structure has been taken up to collapse under the first load (in any loading cycle) the moment diagram has two hinges and the collapse load is

$$\tfrac{3}{2}\mu = \frac{PL}{4} \Rightarrow P = 6\mu/L$$

As the structure unloads elastically the residual support moment can be computed as

$$\mu - \frac{3}{32}PL = \mu - \frac{3}{32}6\mu = \frac{7}{16}\mu$$

Similarly, for the second load case, as the structure unloads from the collapse load, the residual moment can be computed as

$$-\mu + \tfrac{3}{16}PL = -\mu + \tfrac{3}{16}6\mu = \mu/8$$

At this point the residual moment diagrams have been computed for the structure after each load set has been removed.

Incremental collapse can be argued in the following manner. As the first load is applied, a hinge forms under the load and an inelastic discontinuity develops there. If the second load is applied a hinge and a discontinuity develop at the center support. Because of the residual moments left by load set 2, when load set 1 is again applied there is again inelastic defomration at the hinge under the load. In this manner it can be argued that every time either load is applied and removed there is an increment of rotation added to one of the hinges. After many load cycles these increments add up and imply incremental collapse.

5.7.4 Common Sense

Fig. 5.16 makes a point about effective stiffness. A plastic analysis of both beams shown in Fig. 5.16 gives the same collapse load

$$P = 6\mu/L$$

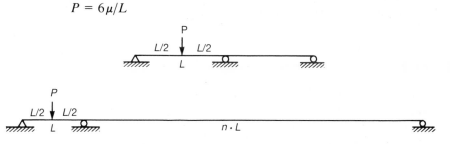

Long side span reduced the effective response of adjacent span

Fig. 5.16

This is 50% above the collapse load of a simply supported beam and illustrates the beneficial effect of continuity. That is, there is a 50% gain in strength achieved by making this beam continuous over the center support. The problem is that as far as plastic analysis is concerned, both of the structures in Fig. 5.16 are identical. But from an elastic point of view, the center joint becomes more and more flexible as the side span becomes longer and longer. Practically, if the side span gets long enough it offers negligible support to the other span which then appears to be simply supported. When a more careful analysis is performed, the effect of the side span flexibility is seen quite clearly as leading to large deflections before the joint restraint comes into play. Plastic analysis simply neglects this fact.

5.8 EXERCISES

1 *Wide flange sections*. The most general commonly used steel beams are called wide flange sections. Available wide flange beams are described in the *AISC Manual* where their properties are tabulated. In fact, the properties of one particular case were used in Section 5.2 above.

   Since these sections are produced by rolling steel ingots, they are formed with rounds and fillets which make computations of section properties more difficult. In order to get an idea of what is going on these beams can be approximated by rectangular elements. For example

W 18 x 55

From the *AISC Manual*:
Depth – 18.11 in
Web thickness – 0.39 in
Flange width – 7.53 in
Flange thickness – 0.63 in
Area – 16.2 in^2
I – 890 in^4
Z – 112 in^3

Approximately:

Approximately:

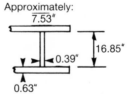

Assume the web and flanges to be made from rectangular plates.

The moment of inertia I for the approximate section is

$$I = \frac{0.39 \times 16.83^3}{12} + 2\left(7.53 \times \frac{0.63^3}{12} + 0.63 \times 7.53 \times 8.74^2\right)$$

$$= 880.1 \text{ in}^4 \quad \text{(compare with 890 for the real section)}$$

Using the rectangular stress distribution associated with a plastic hinge, the ultimate moment for the approximate section is

$$M_{\text{ult}} = f_y \times 2\left[7.53 \times 0.63 \times 8.74 + 8.42 \times 0.39 \times \frac{8.42}{2}\right]$$

$$= f_y \times 110.6$$

The coefficient of f_y in this expression is called the 'plastic section modulus' z. Therefore

$$z = 110.6 \text{ in}^3 \qquad \text{(compare with 112 for the real section)}$$

Problem: A structural tee is formed by splitting a wide flange section in half. Find I and Z for a tee cut from a $W\,18 \times 55$ beam.

2 *Plastic analysis of beams.* There is a decomposition of beam moment diagrams which can be useful in plastic analysis. Simply put it involves

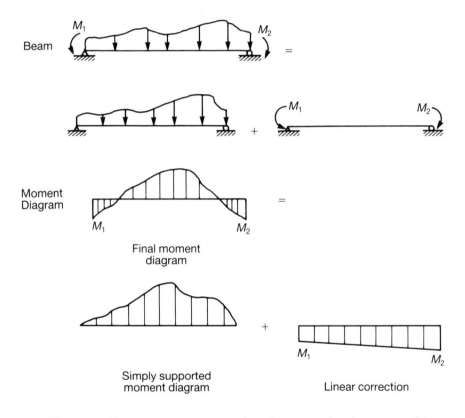

The most direct way to argue such a decompsotion is superposition: since the equations of equilibrium are linear, loads can be dealt with separately and their effects subsequently added. Some applications of this decomposition are indicated below.

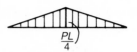

Simply supported
moment diagram

$$PL/4 = 2\mu \Rightarrow P = 8\mu/L$$

(b)

Beam

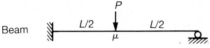

Beam when
simply supported

Collapse
moments

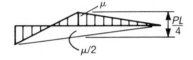

Simply supported
moment diagram

$$3/2\mu = PL/4 \qquad = \qquad P = 6\mu/L$$

(c)

Beam

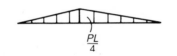

Beam when
Simply supported

Collapse
moments

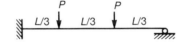

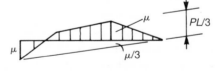

Simply supported
moment diagram

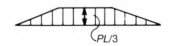

$$\mu + \mu/3 = PL/3 \Rightarrow P = 4\mu/L$$

Problem: Find the collapse loads for the beams shown below.

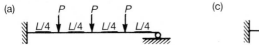

(a)

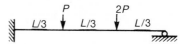

(c)

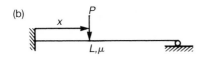

(b)

3 Find the collapse load using the upper bound theorem.

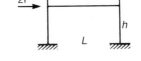

columns – μ_c
beams   – $\mu_b = 2\mu_c$

4 Plot the collapse load as a function of the parameter α.

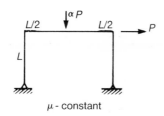

μ - constant

5 Find the collapse load.

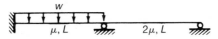

6 Find the collapse load.

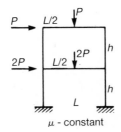

μ - constant

7 Find the collapse loads.

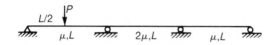

8

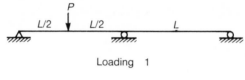

9 Discuss shakedown of the following problem.

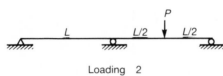

10 Find the collapse load.

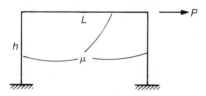

11 Find the plastic collapse load P.

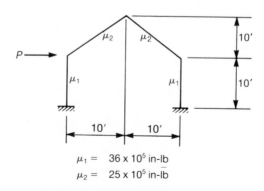

$$\mu_1 = 36 \times 10^5 \text{ in-lb}$$
$$\mu_2 = 25 \times 10^5 \text{ in-lb}$$

Solution —

Rather than solve this problem in its entirety, a collapse mechanism will be assumed and then proved to be the true collapse mechanism.

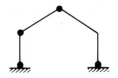

Kinematics:

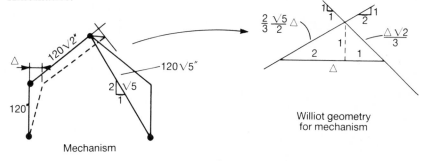

Williot geometry
for mechanism

Mechanism

Work energy equation:

$$P\Delta = \frac{\Delta}{120}(36 + 25) \times 10^5 + \frac{\Delta}{3}\frac{\sqrt{2}}{120\sqrt{2}}(25 + 25) \times 10^5$$

$$+ \frac{\Delta\sqrt{5}}{3}\frac{1}{120\sqrt{5}}(36 + 25) \times 10^5$$

$$\underline{P = 81.5 \times 10^3\,\text{lb}}\qquad \text{(upper bound on the collapse load)}$$

Static approach:

Use the 81.5^K load and construct a safe and statically admissible moment diagram.

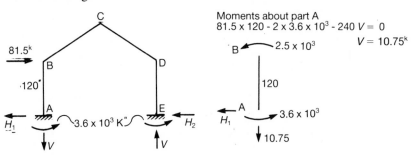

Moments about part A
$81.5 \times 120 - 2 \times 3.6 \times 10^3 - 240\,V = 0$
$V = 10.75^k$

$$M_D = 30.7 \times 120 - 3.6 \times 10^3 = \underline{84 \text{ K}''} \qquad (M_D \text{ safe})$$

$$M_C = 30.7 \times 240 - 10.75 \times 120 - 3.6 \times 10^3$$
$$= 2475 \text{ K}'' \qquad \text{(should be 2500) problems with round-off}$$

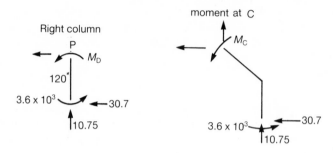

Problem:

Show that this is not the true
collapse mechanism

CHAPTER 6

Uses of Cables

Many of the interesting developments in structural engineering in the past thirty years have involved the use of cables. Following the Second World War prestressed concrete, which combines cables with concrete beams, became a common type of construction (prestressed steel is also used); again following the Second World War cable-stayed bridges, because of their ease of construction, became more and more common; and recently air-supported, fabric-covered cable nets (see the photograph of the Osaka Pavilion) have appeared as inexpensive means of covering large spaces.

From the point of view of design, cables make the most effective use of structural material. For example, while the allowable tensile stress used in mild steel is in the range of 20 000 p.s.i., the allowable tensile stress used in cables can be well over 100 000 p.s.i. Beyond this cable structures carry their load through tension which turns out to be the most efficient way of doing so and much more efficient than bending – given the choice. And of course the longest span bridges are suspension bridges.

Cables tend to add a new dimension to this book. Thus far it has been assumed that deflections (and rotations) are small and that loads do not change the shape of a given structure.

In the case of cables, the loads actually determine the shape of the structure. Fig. 6.1 illustrates some of the effects that are characteristics of cable structures. First, note that the two collinear bars used form a geometrically unstable structure in terms of the material of Chapter 2. This is manifest physically by the fact that when this structure is built it appears 'floppy' or that the approximate member/joint displacement used in Chapter 3

$$\Delta_i = \mathbf{n}_i \cdot (\boldsymbol{\delta}_A - \boldsymbol{\delta}_c) \tag{6.1}$$

implies no member length change under a vertical displacement of the center point. On the other hand, if such a structure is prestressed (think of a taut string) it can work perfectly well as a load-carrying member. In fact, for small angle changes and constant tension T – which is consistent with the remark above concerning the absence of member length change –

US Pavilion – Expo '70, Osaka, Japan. (Courtesy of David Geiger, Geiger/Berger Engr, NY City.)

there is a linear relationship between the applied load and the displacement as indicated in Fig. 6.1.

It is then prestress which makes the structure of Fig. 6.1 work. The

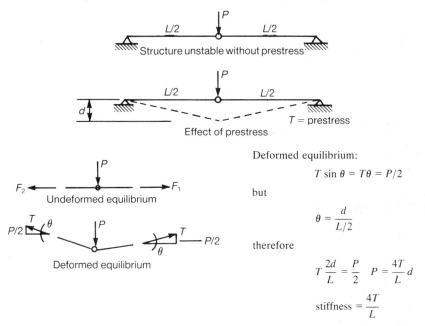

Fig. 6.1 – Effect of prestress on a two-bar truss

fact that prestressed effects have become primary load-carrying mechanisms of the structure is explained by the theorem (not proven here) which
states that

In the presence of prestress, first order non-linearities can be of the
same order as the classical linear effects.

This tells the engineer that when prestress is present, linear analysis may
not be adequate.

This chapter attempts to present an overview of cables and their
applications. Unfortunately, a thorough discussion of some very interesting
topics such as the layout and analysis of cable nets is beyond the scope
of this text.

6.1 PLANE CABLES

In this section the differential equation of a cable loaded vertically in a
plane is discussed together with several elementary solutions. In one way

or another, this differential equation is responsible for most of the common applications of cables including prestressed concrete and to some extent cable nets.

As is common in mechanics, a short section of cable will be examined in order to derive a differential equation which characterizes behavior in the large cable. (See Fig. 6.2.) Since the load is vertical, it is usual to

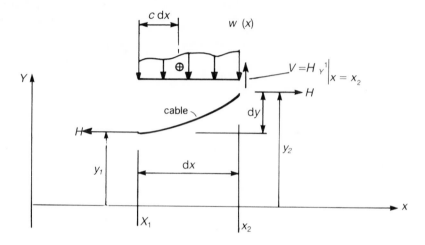

Fig. 6.2 – A cable element

single out the horizontal projection H of the cable tension; H must, of course, remain constant when the load is vertical.

First, it should be noted that like the truss bar, moment equilibrium for a cable requires the resultant force in the cable to be tangent to the cable itself. The distributed load has a secondary effect in the moment equation. This can be shown by taking moments about, for example, the point $x = x_1, y = y_1$ which gives

$$wc(dx)^2 + H\,dy - V\,dx = 0$$

or

$$\frac{V}{H} = \frac{dy}{dx} = y'$$

neglecting higher order terms. This motivates writing the vertical force component V as

$$V = Hy'$$

Summing forces in the vertical direction now results in

$$w \, dx + Hy' \big|_{x=x_1} = Hy' \big|_{x=x_2}$$

or

$$w = H \frac{y' \big|_{x=x_2} - y' \big|_{x=x_1}}{dx} \Rightarrow \quad w = Hy'' \tag{6.2}$$

Eq. (6.2) is the basic differential equation of a cable subjected to a vertical load in a plane. Special solutions of this equation which are commonly used will now be investigated.

6.1.1 Particular Solutions

Case 1. $w(x) = \bar{w}$ constant
One of the most common results of cable theory is that a cable under a uniform load in the horizontal direction takes on a parabolic shape. (Note that this is not the case of a cable under its own weight!) This result can be shown simply by integrating Eq. (6.2) as

$$\begin{aligned}
Hy'' &= \bar{w} \\
Hy' &= \bar{w}x + c_1 \\
Hy &= wx^2/2 + c_1 x + c_2
\end{aligned} \tag{6.3}$$

where c_1 and c_2 are constants of integration.

Case 2. A Cable with a Kink
It will be shown that a kink in a cable inplies a concentrated force and vice versa. This follows directly from a free body diagram of a cable with a kink (Fig. 6.3). In this case equilibrium implies

$$F = H \frac{b+c}{a} \tag{6.4}$$

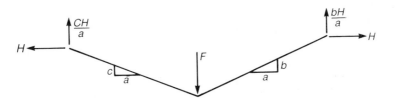

Fig. 6.3 – Free body diagram of a cable with a kink in it

Case 3. A Cable Under its own Weight (the Catenary)
A cable loaded only by its own weight is subjected to a load $w(x)$

proportional to its arc length. (This is not Case 1 above.) Using the well known relationship

$$ds^2 = dx^2 + dy^2 \Rightarrow ds = dx\sqrt{1 + (y')^2}$$

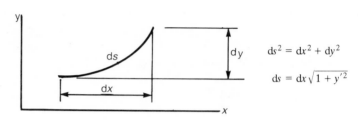

Fig. 6.4 – Arc length

(see Fig. 6.4), if a cable weighs w_s lb/ft, the horizontal distribution of load becomes simply

$$w = w_s\sqrt{1 + (y')^2} \Rightarrow Hy'' = w_s\sqrt{1 + (y')^2} \tag{6.5}$$

This equation can be integrated as follows. Let $y' = p$. Eq. (6.5) then becomes

$$w_s\sqrt{1 + p^2} = Hp' \Rightarrow \frac{dp}{\sqrt{1 + p^2}} = \frac{w_s}{H}dx$$

Integrating

$$\sinh^{-1}p = \frac{w_s}{H}x + c_1 \Rightarrow \frac{dy}{dx} = \sinh\left(\frac{w_s}{H}x + c_1\right)$$

$$y = \frac{H}{w_s}\cosh\left(\frac{w_s}{H}x + c_1\right) + c_2$$

c_1 and c_2 are again constants of integration.

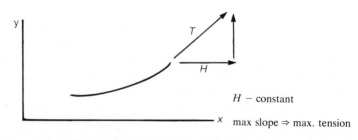

H – constant

max slope $\Rightarrow$ max. tension

Fig. 6.5 – Cable tension

These three cases are the most common load distributions encountered in practice. In fact, it will be seen below that the parabola is a good approximation of the catenary when the sag is small. For reasons of simplicity it is therefore somewhat uncommon to deal directly with the catenary in most applications.

Before discussing specific cases it is worthwhile to note that since the loading is vertical and the horizontal component of the cable force H is constant, the tension varies from point to point as a function of the cable slope (Fig. 6.5) and is maximum where the slope is maximum. This follows directly from the fact that the resultant cable force is parallel to the cable slope:

$$T = H\sqrt{1 + (y')^2} \quad \text{or} \quad T = H \sec \theta$$

Finally, note that for the catenary, the change in cable shear between two points is proportional to the cable length between these points (Fig. 6.6). That is

$$w_x s + Hy'(x_1) = Hy'(x_2)$$

or

$$s = \frac{H}{w_s}[y'(x_2) - y'(x_1)]$$

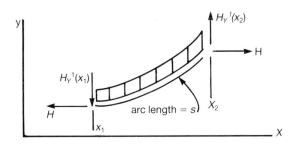

Fig. 6.6 – Cable length ~ Change in shear

6.1.2 Some Examples: the Parabola versus the Catenary

The example of Fig. 6.7 is useful because it is one of the few cases in which it is possible to compare exactly the catenary with its parabolic approximation. As indicated above, the argument is that when the sag is small the parabola is a good approximation of the catenary.

First, note that for moment equilibrium of the tower itself (see Fig. 6.7) the force component H must be 10^K. It then remains to use the two boundary conditions to solve for the two constants of integrations in the cable equation. Consider the 'exact' case in which the cable is regarded to be a catenary. In view of the discussion above it is necessary to solve the expression

$$y = \frac{H}{w_s} \cosh\left(\frac{w_s}{H} x + c_1\right) + c_2$$

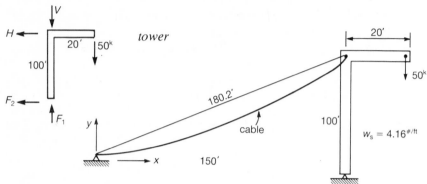

Free body diagram of derrick

Moments about base: $50 \times 20 = 100H \Rightarrow H = 10^K$

Fig. 6.7 – Cable supported derrick

for the constants c_1 and c_2. Using the fact that $y = 0$ at $x =$ it follows that

$$0 = \frac{H}{w_s} \cosh(c_1) + c_2$$

The secondary boundary condition, $y = 100$ at $x = 150$, implies that

$$100 = \frac{H}{w_s} \cosh\left(\frac{w_s}{H} \times 150 + c_1\right) - \frac{H}{w_s} \cosh c_1$$

Using Appendix 8 this transcendental equation can be solved as

$$c_1 = b = \sinh^{-1}\left\{\frac{c}{2\sinh(a/2)}\right\} - \frac{a}{2}$$

where

$$H = 10^K, \qquad w_s = 4.16 \text{ lb/ft}, \qquad c = \frac{4.16}{10\,000}$$

$$a = \frac{w_s}{H} 150 = \frac{4.16}{10\,000} \times 150 = 0.0624$$

or

$$c_1 = b = 0.5945$$

It follows that

$$y' = \sinh\left(\frac{w_s x}{H} + c_1\right) = 0.705 \quad \text{at} \quad x = 150'$$

and

$$T_{max} = 10 \sec \theta = 10\sqrt{1 + (0.705)^2} = 12.23^K$$

Consider now a parabolic approximation of this exact solution. To make such an approximation requires the determination of an approximate uniformly distributed horizontal load w which is to be associated with the parabolic shape. In view of the data available, the best that can be done is to approximate the (unknown) exact cable length by the cable secant as indicated in the figure and correct the weight as

$$w = \frac{180.2}{150} \times 4.16 = 5 \text{ lb/ft}$$

It now remains to solve (see Case 1 above)

$$Hy = wx^2/2 + c_1 x + c_2$$

for the constants c_1 and c_2. Since $y = 0$ at $x = 0$, $c_2 = 0$; since $y = 100'$ at $x = 150$, it follows that

$$10 \times 100 = \frac{0.005 \times 150^2}{2} + c_1 \times 150$$

or

$$c_1 = \frac{943.7}{150} = 6.3$$

Again the cable slope is maximum at the right end and therefore the tension is also maximum there.

$$Hy' = wx + c_1$$

$$\tan \theta = y' = \frac{wx + c_1}{H} = \frac{0.005 \times 150 + 6.3}{10} = 0.704$$

$$T = H \sec \theta = 12.3^K$$

With an eye to engineering accuracy, the parabola and the catenary should be regarded as producing equivalent results in this case.

6.1.3 A Cable Loaded by Concentrated Forces

Fig. 6.8 shows a cable loaded by concentrated forces whose horizontal locations are fixed but whose vertical positions remain to be determined. This is of course similar to the case of a cable under continuous loading $w(x)$ whose position $y(x)$ must be determined as part of the solution. The problem can be posed in several ways. Most simply, if the horizontal cable force component H is also specified, then the reactions R_1 and R_2 can be computed using moment equations as is done in the case of beam reactions. Alternatively, some vertical location along the cable (or the sag) can be specified and then H, R_1, and R_2 computed. In both cases, given these force components it is possible to compute the node locations along the cable.

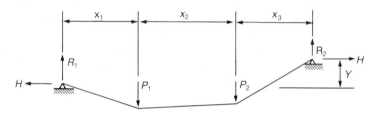

Fig. 6.8 – Cable subjected to point loads

For example, R_1 and H imply the slope of the first cable segment, $R_1 - P_1$ and H imply the slope of the second segment, etc. In this manner all the points along the cable can be located.

Example 6.1 discusses these alternatives. In the first case the horizontal force component H is assumed to be given to be 10^K. R_2 can be shown to be 25^K using a moment equation about the left support. This implies $R_1 = 45 - 25 = 20^K$. Finally, node locations are computed along the cable using known cable slopes.

Example 6.1   Cable subjected to point loads

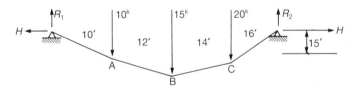

Case 1. H specified, say $H = 10^K$

Moment about left support $= 0$ $52 R_2 = 10 \times 10 + 15 \times 22$

$52 R_2 = 10 \times 10 + 15 \times 22 + 20 \times 36 + 10 \times 15$

$R_2 = 25 \qquad R_1 = 45 - 25 = 20$

Slope at left segment $= 20/10 \Rightarrow y$ coord $A = -20'$

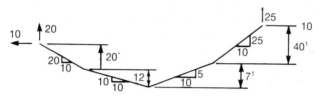

Case 2.   Vertical location of point B fixed

Say point B $20'$ below left support

Moments about point B $= 0$

$20 + 14 - 30 R_2 + 35 H = 0$

Moments about left support $= 0$

$52 R_2 = 10 \times 10 + 15 \times 22 + 20 \times 36 + 15 H$

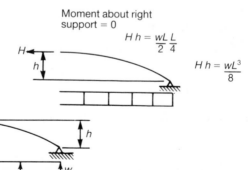

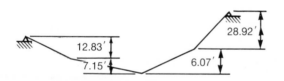

two equations $-$ two unknowns

$H = 14.56^K$

$R_2 = 26.52 \Rightarrow R_1 = 18.68$

Case 3.   Continuous loading

Moment about right
support $= 0$

$$H h = \frac{wL}{2} \frac{L}{4}$$

$$H h = \frac{wL^3}{8}$$

If, for example, it is determined that the cable has too much sag and that it is desirable to fix the low point at $20'$ below the left support, this implies a relationship between H and R_2 which can be obtained using a free body diagram of the cable cut at this low point. This equation together with a moment equation for the entire structure about the left support provide two equations to determine R_2 and H. Finally, vertical node locations are computed as in Case 1 above.

Note that in the above examples, H was increased to reduce the sag. This trade-off between sag and cable tension is common. For example, Case 3 of Example 6.1 might be thought of as a section of an air-supported roof or a single parabolic cable. From the free body diagram of half of the structure it follows that

$$Hh = wL^2/8$$

or that the product of H times h is constant. Clearly increasing H reduces the rise h and vice versa.

6.2 BEAMS WITH CABLES

Elementary beam theory implies that a beam in flexure will experience both tensile and compressive stresses on any cross-section. When this is applied to concrete beams difficulties arise immediately because concrete has very little tensile strength. A plain concrete beam would then simply crack in brittle fracture at a very low load. To avoid tensile cracking, steel reinforcing bars are added to 'carry the load' in areas of tension in a beam cross-section. What results is 'reinforced concrete'.

Another way to deal with the tensile stresses associated with flexure in concrete is to start with a state of compression. If the compression is large anough, the superimposed tensile stresses of bending will never produce a resultant tension thus avoiding tensile cracking. One way to produce this initial compression is through the use of cables. For example, if a prestressing cable is run along the centroid of a beam it will produce an initial state of constant compression. Generally, more complex cable configurations are used.

Whatever the beam/cable configuration, there is an analysis to be performed which is the topic of this section. As far as the concrete beam itself is concerned, it is subjected to two sets of loads: the external loads and the load exerted upon it by the prestressing cable or cables. In order to design the concrete beam itself it is first necessary to determine the loads exerted on it by its prestressing cables. After that it is simply a question of determining shear, moment and thrust diagrams as usual.

Example 6.2 illustrates this type of analysis. The first step is to construct a free body diagram of the cable itself. Given the cable shape and the prestressing, the forces required to keep the cable in equilibrium can be

Example 6.2   Draw the stress resultant diagrams for the prestressed concrete beam shown.

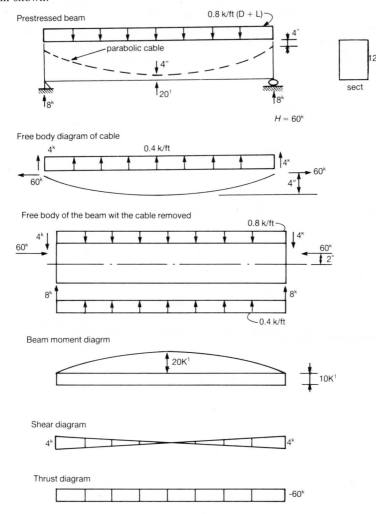

determined. In this case the parabolic cable implies that

$$wL^2/8 = hH \Rightarrow w = \frac{Hh8}{L^2} = \frac{60 \times \frac{1}{3} \times 8}{20^2} = 0.4 \text{ K/ft}$$

(see Fig. 6.8).

Given the forces on the cable in this example, when the cable is removed from the beam, equal and opposite forces must act on the concrete beam. These forces then complete the free body diagram of the concrete beam. Stress resultant diagrams follow directly. Note that in this case the cable carries half the applied load. If H were soubled the cable would carry all the applied load but the beam would still be subjected to a bending moment due to the fact that the cable terminates above the beam centroid at the ends of the beam. Clearly, by terminating the cable at the centroid a completely 'balanced' beam could be designed so that under load the beam would only see a thrust ($V = M = 0$). There are many creative possibilities for the engineer within prestressed concrete.

Example 6.3 describes a prestressed concrete beam whose cable is kinked. It is similar to Example 6.2 and requires no further discussion.

Example 6.3 Draw the stress resultant diagrams for the prestressed concrete beam shown.

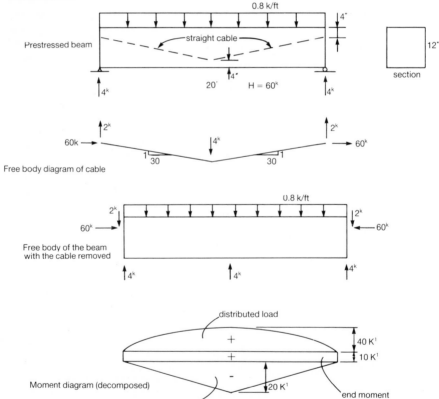

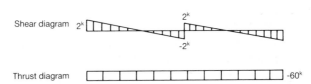

Moment diagram (combined)

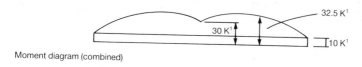

6.3 A STATICALLY DETERMINATE SUSPENSION BRIDGE

Fig. 6.9 shows a suspension bridge with a three-hinged stiffening girder and a unit load applied at some distance ξ from the left end. It will be seen that under rather stringent assumptions this structure is statically determinate.

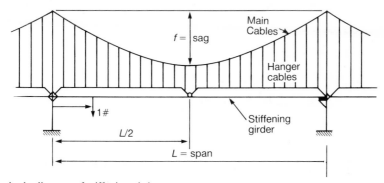

Free body diagram of stiffening girder

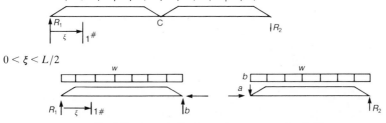

Fig. 6.9 – A statically determinate suspension bridge

ΣM_c on left $\qquad R_1 \dfrac{L}{2} + \dfrac{wL^2}{8} - (L/2 - \xi) = 0$

ΣM_c on right $\qquad R_2 \dfrac{L}{2} + \dfrac{wL^2}{8} = 0$

$\Sigma F_{\text{vert}} \qquad\qquad R_1 + R_2 + wL = 1$

Adding:

$$(R_1 + R_2)L/2 + \frac{wL^2}{4} = L/2 - \xi$$

$$(R_1 + R_2 + wL)L/2 + \frac{wL^2}{4} = L/2 - \xi + \frac{wL^2}{4}$$

$$L/2 + \frac{wL^2}{4} = L/2 - \xi + \frac{wL^2}{4} \Rightarrow w = \frac{4\xi}{L^2}$$

While the analysis which follows is far too simple to be applied in practical design situations, it does give some idea of how suspension bridges work. In this analysis it is assumed that the hangers and the stiffening girder combine to keep the main cables in their initial parabolic configuration. Furthermore, it is assumed that the hangers exert a uniformly distributed load w on the stiffening girder. It follows that the horizontal force component H is related to w as before through (see Fig. 6.8)

$$\frac{wL^2}{8} = Hf$$

A free body diagram of the stiffening girder can finally be used to compute w given the load position ξ (see Fig. 6.9).

$$w = 4\xi/L^2$$

6.4 A SIMPLE EXAMPLE OF AN AIR-SUPPORTED STRUCTURE

Air-supported structures comprise one of the most interesting structural engineering developments of the 1970s. Those like the Osaka Pavilion (see photograph at beginning of chapter) are particularly simple with three basic elements: a fabric covering, a cable net, and a circumferential grade beam. Usually, the covering is Teflon-coated fiberglass held up by a small internal pressure. It is also common to select the layout of the structure so that the circumferential grade beam is funicular (has no bending moment) under some particular loading condition.

A cable layout for symmetric structures like the Osaka Pavilion can be developed in the following manner (see Fig. 6.10):

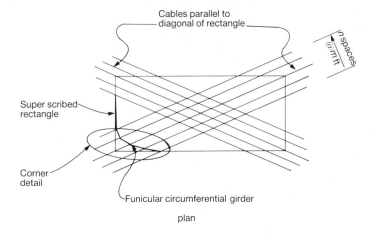

Fig. 6.10 – Cable net layout

Step 1.   Select the dimensions of the superscribed rectangle.

Step 2.   Select the number of cables and their spacing in plan. Lay out these cables parallel to the diagonals of the superscribed rectangle.

Step 3.   Solve the corner problem for a funicular circumferential girder (see Fig. 6.11). This can be done by *assuming* the values of the horizontal projections of the cable forces under the restriction that their resultant passes through the intersection of T and T'. Given these projections the corner layout problem becomes a problem of a cable under concentrated loads (see Section 6.1.3) which can be solved for the locations of the points

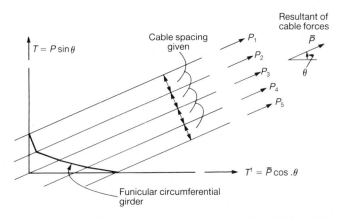

Notes:
(1) The cable resultant force ^-P must pass through the intersection of T' and T.
(2) Finding the circumferential girder coordinates then becomes a cable problem under point loads.

Fig. 6.11 – Corner detail

Example 6.4   Cable net layout.

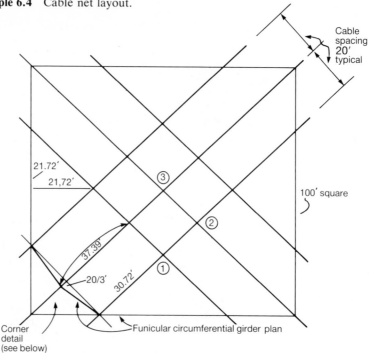

Assumptions:
(1) Internal pressure 4 p.s.f.
(2) Horizontal projection of cable force at corner 4^K.
(3) Superscribed square $100' \times 100'$.
(4) Cables spaced at $20'$.
(5) Only consider the vertical load component of internal pressure.

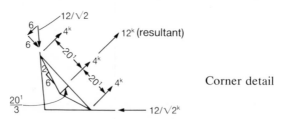

Corner detail

Vertical loads due to internal pressure:

Point	Area	Load (lb)
3	$20 \times 20 = 400 \text{ ft}^2$	1600
2	$10 \times 20 + 20 \times 37.39/2 = 574 \text{ ft}^2$	2296
1	$10 \times 10 + 10 \times 30.72/2 \times 2 + 21.72\frac{2}{3} = 564$	2256

Center point $-$ 3 (see plan)

Symmetry requires that each cable have a vertical force component of $1600/4 = 400$ lb

> therefore cable slope $= 400/4000 = 0.1$ in this section.

Use vertical equilibrium to determine the heights at points 2 and 1, h_2 and h_1

$$\text{point 2} \quad 4000 \left(\frac{h_2}{37.39} + 2 \frac{h_2 - h_1}{20} \right) = 2296 + 400 = 2696$$

$$\text{point 1} \quad 2 \times 4000 \times \frac{h_1}{30.72} = 2256 + \frac{2 \times 4000}{20}(h_2 - h_1)$$

Solve for h_1 and h_2: $h_1 = 0.73'$ $h_2 = 5.89'$

of attachment of the cables to the circumferential girder. Note that these assumptions concerning the horizontal force projections eventually fix the elevations of the cable intersections.

Step 4.   Vertical equilibrium can finally be used to determine the heights of the cable intersections above the circumferential girder.

Example 6.4 illustrates these steps for a simple cable net configuration. In it the distributed forces due to internal pressure within the cable net are approximated by concentrated forces at the cable intersections.

6.5 CABLES IN THREE DIMENSIONS

Example 2.21 gives the three equations of equilibrium of a three-dimensional cable as a special case of the equilibrium of a three-dimensional beam. These are:

$$\mathbf{P}' + \mathbf{p} = 0 \Rightarrow \quad P'_t + p_t = 0$$
$$P_t/\rho + p_n = 0$$
$$p_b = 0$$

It should be noted that the approach used for these equations is quite different from that of this chapter. The differential equation $w = Hy''$ is really written in a global coordinate system while the above equations move with the coordinates of the cable. If these coordinates are not known *a priori* it becomes difficult to use this latter form.

To illustrate the use of these equations, there is a simple solution obtained when a helical cable with helix angle α is wrapped around a smooth cylinder of radius R. In this case P_t is constant, $p_n = -P_t/\rho = -P_t R/\cos^2 \alpha$ (see Section 2.3.3.1) and $p_b = 0$.

6.6 EXERCISES

1 For the prestressed concrete beam shown, compute and plot the moment, shear, and thrust diagrams (about the centroid). The beam is rectangular with a width of $12''$. The cable is segmentally straight with a tension of 100^K. Do not neglect the dead load.

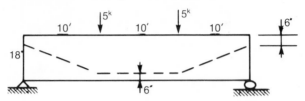

2 The drawing shown represents part of a cable-supported roof. The cable weighs 10 lb per linear ft. Compute the maximum tensile force in the cable assuming the cable to be: (a) parabolic, (b) catenary.

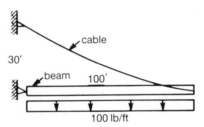

3 Compute and plot the moment at the left quarter point in the stiffening girder of a three-hinged, symmetric suspension bridge with span L, cable sag f for a unit load at any point (x) on the span.

4 For $H = 10^K$, compute the equilibrium position of the cable shown (neglect the weight of the cable).

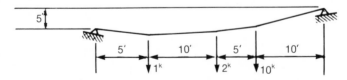

5 A cable is used to support a load P at the centre of a $1000'$ span. The cable is a symmetric and has a maximum sag of $400'$. If the cable weighs 10 lb/ft and $P = 10^K$, find the maximum tension in cable considering the shape of the cable to be (a) a catenary (exact), (b) a parabola (approx.).

6 Draw the moment, shear, and thrust diagrams (at the centroid) for a free body of the concrete beam shown. The beam carries a load of

1.6 k/ft (total load including its own weight) span = 40', beam web and flanges are 4″ thick, the prestressing cable has a parabolic shape ($H = 152^K$).

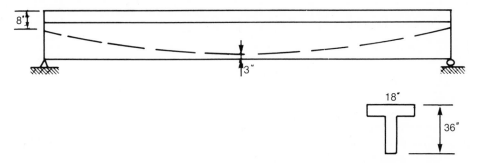

7 Redo the calculations of Example 6.4 using two cables spaced at 30' parallel to each diagonal.

8 Discuss the modifications which must be made in Example 6.4 to account for horizontal force components of internal pressure applied to cable intersections.

CHAPTER 7

Moment distribution

Moment distribution is an iterative method of structural analysis developed by the late Professor Hardy Cross in about 1930. It had a major impact upon the structural engineering profession and was taught as one of the principal methods of structural analysis through the 1950s and 1960s but has been displaced more recently by computer-based methods.

The reader will by now be aware that structural analysis can require considerable effort on the part of the engineer when no computers are available. In fact, the choice of examples suitable for this text is severely limited by the computational effort they require of the student. Faced in the past with structural analysis without the computer, the engineer had either to solve large systems by hand or look to approximate methods. Moment distribution appeared as a breath of fresh air in the 1930s since it placed a new class of problems well within the engineer's grasp at modest computational effort.

This chapter will discuss some elementary aspects of moment distribution. It will be described as an iterative approach to the displacement method in which constraints are introduced and then relaxed iteratively. In these terms moment distribution can be thought of as a general method of analysis. Unfortunately, as a general method it is ineffective and can be difficult to execute. There are really two reasons for discussing moment distribution here. In the first place, there is a class of beam and frame problems for which moment distribution can still be an effective method of analysis. But more importantly, moment distribution is an extremely physical method of analysis. The engineer cannot perform moment distribution without developing insight into how his structure works.

Finally, the reader should note that as commonly practiced and as presented here, moment distribution neglects member length change.

7.1 MEMBER STIFFNESS AND JOINT DISTRIBUTION

Basic to moment distribution is the idea of member stiffness. (See Fig. 7.1.) This is the moment required to produce a unit rotation at one end of a beam when the other end is fixed. It was shown earlier in Chapter 3 that for a uniform beam this moment is $4EI/L$. Fig. 7.1 also makes the point that member stiffness can be defined for members with variable moment of inertia. In that case the stiffness of one end of a beam is not necessarily equal to the stiffness of the other end. On the basis of Fig. 7.1 the carry-over factor is also defined. Because of the linearity of the beam equations,

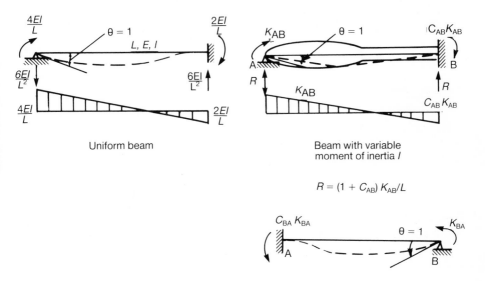

$$R = (1 + C_{AB}) K_{AB}/L$$

Fig. 7.1 – Member stiffness. Member carry-over

the solution of Fig. 7.1 can be scaled for any joint rotation. In doing so the ratio of the moment at the left end to the moment at the right end remains at half. Thinking of the applied moment at the left end as producing the moment at the right end, it is said that the carry-over factor for a uniform bean is half. Carry-over factors are also defined for beams with variable moment of inertia but they may not be half and they may not be the same at both ends.

Fig. 7.2 indicates a basic problem of moment distribution which can now be solved easily. The question is, given a unit moment applied to joint A, what moments are produced in each of the members. One way to proceed is to solve the problem for a unit joint rotation and then scale the resulting solution for a unit moment. For a unit joint rotation the moment in each member at joint A is just its stiffness $K_i = 4EI/L_i$. The applied moment must then be

$$M = \sum_{\text{members}} K_i$$

which is called the joint stiffness. For an applied unit moment this solution scales so that the moment in each member at joint A is

$$K_i \bigg/ \sum_{\text{members}} K_j$$

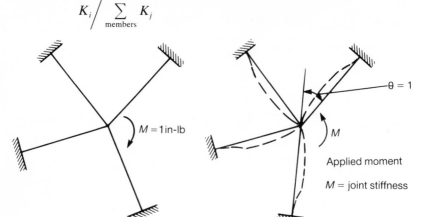

Fig. 7.2 – Joint distribution

Example 7.1

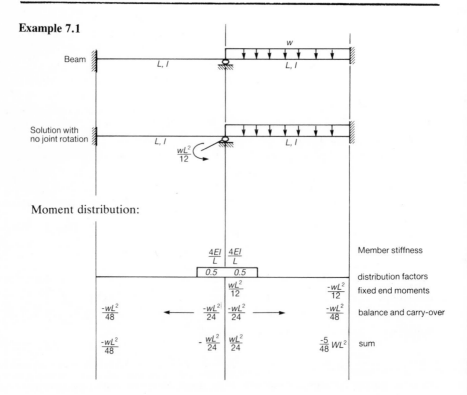

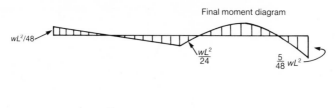

Final moment diagram

$wL^2/48$

$\dfrac{wL^2}{24}$

$\dfrac{5}{48}wL^2$

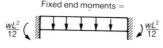

Fixed end moments =

$\dfrac{wL^2}{12}$ $\dfrac{wL^2}{12}$

This moment is called the distribution factor of member i at joint A. That is

$$\text{member distribution factor} = \frac{\text{member stiffness}}{\text{joint stiffness}}$$

It is now possible to solve the problem of Example 7.1. First, the center joint is fixed (the rotation is set to zero) which gives the so-called fixed-end moment solution for the beam on the right and no response in the beam on the left. As in the case of the displacement method of Chapter 4, this solution is valid except that it requires an external moment to be applied to the center joint. The final solution is constructed by 'releasing' or 'balancing' the center joint which is equivalent to applying a clockwise moment of $wL^2/12$ to this joint. Using the ideas of distribution and carry-over the solution is completed in this figure. Note that the sign convention used implies that a counter-clockwise moment on the end of a member is positive. Care must then be exercised in drawing the final moment diagram – which uses a different sign convention.

Example 7.1 is too simple to show the manner in which repeated iteration is generally required in moment distribution. To do so consider a modified version of this example indicated in Example 7.2. In this case the support on the left has been modified to allow rotation. When moment distribution is applied to this example, the carry-over moment to the left end produces an unbalance there since the joint should be free to rotate. This is handled in the example by balancing the left joint (allowing it to rotate freely). But this, of course, upsets the center joint which must be balanced again, etc. At some point the moment distribution process must be terminated arbitrarily.

The problem of Example 7.2 can also be used to introduce the concept of *modified stiffness*. While the basic moment distribution approach allows one joint to rotate at a time, a more sophisticated approach allows many things to occur simultaneously as long as the basic equations of structures

Example 7.2   Modified version of Example 7.1.

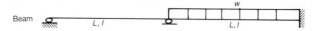

Moment distribution derivation:

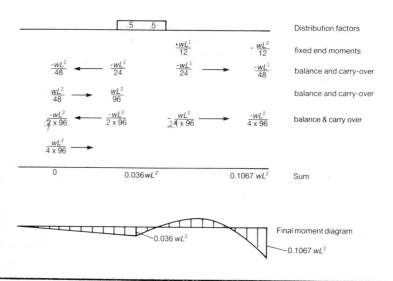

are satisfied at each step. More will be said of this below. In the case of the modified stiffness, when one end of a beam undergoes a unit rotation the other end is allowed to rotate freely. (See Fig. 7.3.) The member stiffness in this case becomes $3EI/L$ for uniform members. This can be

Example 7.3   Example 7.2 re-done using the modified stiffness factor.

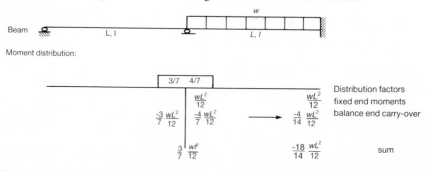

Moment distribution:

determined by simply solving the beam equations or through moment distribution itself as indicated in the figure. Example 7.3 shows how the beam problem discussed earlier can be solved in a single step using the modified stiffness factor.

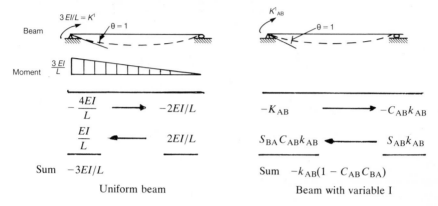

Fig. 7.3 – Modified stiffness factor

7.2 FRAME PROBLEMS AND SIDESWAY

While the discussion of Section 7.1 is adequate for most beam problems, rigid frame problems add a new dimension involving sidesway. When a frame joint is balanced during moment distribution, the structure in general wants to move laterally. If it is not allowed to do so during moment distribution fictitious lateral forces develop which must be accounted for. Put another way, moment distribution ordinarily deals with frames neglecting length change. These structures have a specific number of degrees of freedom: some are rotations and some are displacements. The final solution must accommodate both of these. Doing so requires additional steps since basic moment distribution only deals with joint rotations.

Example 7.4 shows one way of dealing with frames through moment distribution. The logic of the method is described in Fig. 7.4: Joint

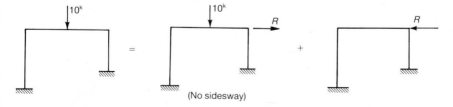

Fig. 7.4 – Sidesway through superposition

Example 7.4   Frame with sideway.

Problem:

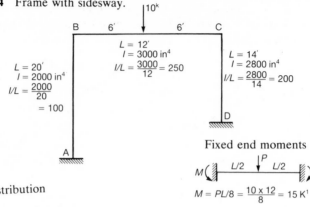

Moment distribution

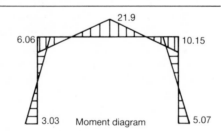

Fixed end moments

$$M = PL/8 = \frac{10 \times 12}{8} = 15 \text{ K}^1$$

Joint	A	B		C		D
member	AB	BA	BC	CB	CD	DC
Distance		0.285	0.715	0.556	0.444	
FEM			15	-15		
	-2.15◄-4.3		-10.7►-5.35			
			5.6◄-11.2	:9.15►4.57		
	-0.8◄-1.6		-4.0 ►-2.0			
			0.55	1.1	0.9◄►0.45	
	-0.08	-0.16	-0.39 ►0.2			
				0.1	0.1◄►0.05	
Total	-3.03	-6.06	6.06	-10.15	10.15	5.07

Moment diagram

Compute horizontal reaction R

$$V_1 = \frac{6.06 + 3.03}{20}$$
$$= 0.454^K$$

column free body diagrams

$$V_2 = \frac{10.15 + 5.07}{14}$$
$$= 1.09^K$$

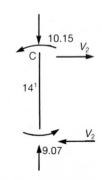

Free body diagram of beam

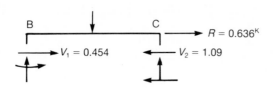

$R = 0.636^K$

$V_1 = 0.454$

$V_2 = 1.09$

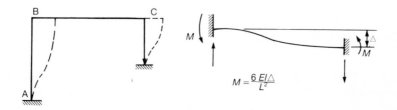

$M = \dfrac{6\,EI\triangle}{L^2}$

Moment distribution

$$\frac{100}{20} = 5 \qquad \text{member AB}$$

FEM's –

$$\frac{200}{14} = 14.3 \qquad \text{member CD}$$

Joint	A	B		C		D
member	AB	BA	BC	CB	CD	DC
Distance		0.285	0.715	0.556	0.644	
FEM	5	5			14.3	14.3
			-4.0← -8.0		-6.3→-3.15	
	-0.142←-0.285	-0.715 ←-0.357				
		0.100←0.199		0.158→0.079		
	-0.014←-0.28	-0.072				
totals	4.844	4.687	-4.687	-8.158	8.158	11.229

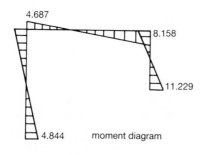

moment diagram

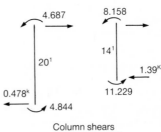

Column shears
total shear = 1.87^k

Superposing solutions:

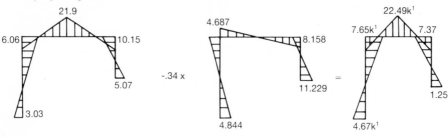

Final moment diagram

balancing within moment distribution naturally removes restraints against joint rotation leaving restraints against joint displacement to be dealt with. That is done in Fig. 7.4 by first constructing a solution which contains the 10^K load. Since this solution has restrained the sidesway, it results in a fictitious restraining force R. A second solution with sidesway is then constructed and eventually scaled so that the fictitious reaction R cancels out when the two solutions are superimposed.

These steps are carried out in detail in Example 7.4. First the 10^K load is applied and the structure is balanced using moment distribution. The fictitious reaction R is then computed using column equilibrium. Next a solution which has sidesway is constructed using fixed-end moments generated by an imposed sidesway without joint rotation. The fictitious restraining force is computed in this case to be 1.87^K. The final solution is constructed as the sum of the first solution plus some constant c times the second solution with c selected so that the fictitious force vanishes,

$$0.636 + 1.87c = 0 \Rightarrow c = -0.34$$

The resulting solution is given in Example 7.4.

This approach to moment distribution with sidesway generalizes readily. If the structure has n sidesway degrees of freedom it is necessary to combine $n + 1$ solutions using n simultaneous equations to eliminate the n fictitious forces which develop.

7.3 GENERALIZATIONS AND EXTENSIONS

There are endless ingenuous variations of moment distribution to be found in the literature. Example 7.5 combines several of these. First, it is noted without proof that

Example 7.5   Symmetric/antisymmetric decomposition.

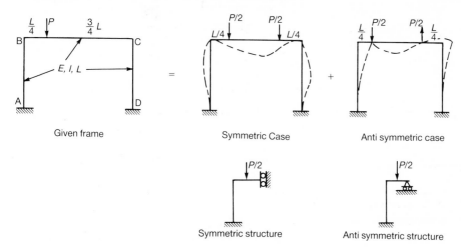

Given frame	Symmetric Case	Anti symmetric case

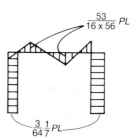

Symmetric structure   Anti symmetric structure

Fixed end moments:

Symmetric solution:

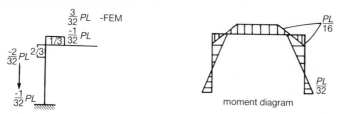

moment diagram

Antisymmetric solution:

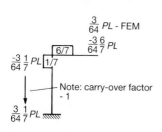

Note: carry-over factor – 1

Final moment diagram:

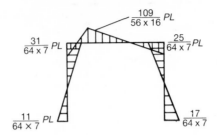

$\frac{31}{64 \times 7}PL$　　　　　　　　　　　　　　　　$\frac{109}{56 \times 16}PL$　　　$\frac{25}{64 \times 7}PL$

$\frac{11}{64 \times 7}PL$　　　　　　　　　　　　　　　　$\frac{17}{64 \times 7}$

An arbitrary loading on a symmetric structure can be decomposed into the sum of a symmetric loading and an antisymmetric loading.

Given such a decomposition several simplifications are possible (see Example 7.5):

The symmetric case. In this case the structure undergoes no lateral displacement and the joint rotations must be equal and of opposite sign.
The antisymmetric case. In this case there is lateral motion of the structure and the joint rotations are equal.

Moments due to lateral displacement:

Moment distribution solution

Uniform beam		Variable I	
$\dfrac{4EI}{L}\dfrac{\Delta}{L} \longrightarrow$	$\dfrac{2EI}{L}\dfrac{\Delta}{L}$	$k_{AB}\dfrac{\Delta}{L} \longrightarrow$	$C_{AB}k_{AB}\dfrac{\Delta}{L}$
$\dfrac{2EI}{L}\dfrac{\Delta}{L} \longleftarrow$	$\dfrac{4EI}{L}\dfrac{\Delta}{L}$	$C_{BA}k_{BA}\dfrac{\Delta}{L} \longleftarrow$	$k_{BA}\dfrac{\Delta}{L}$
$M = \dfrac{6EI\,\Delta}{L^2}$		$(k_{AB} + C_{BA}k_{BA})\dfrac{\Delta}{L}$	$(k_{AB} + C_{AB}k_{AB})\dfrac{\Delta}{L}$

Symmetric stiffness factor:

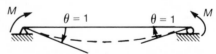

M　　　$\theta = 1$　　　$\theta = 1$　　　M

Uniform beam		Variable I	
$-\dfrac{4EI}{L}$ ⟶	$-\dfrac{2EI}{L}$	$-k_{AB}$ ⟶	$-C_{AB}k_{AB}$
$\dfrac{2EI}{L}$ ⟵	$\dfrac{4EI}{L}$	$C_{AB}k_{BA}$ ⟵	k_{AB}
$M = \dfrac{2EI}{L}$		$-k_{AB} + C_{BA}k_{BA}$	$k_{BA} - C_{AB}k_{AB}$

Antisymmetric stiffness factor:

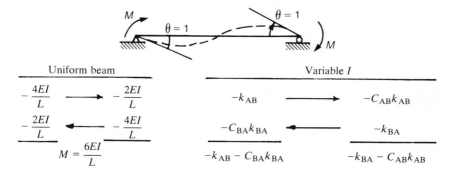

Uniform beam		Variable I	
$-\dfrac{4EI}{L}$ ⟶	$-\dfrac{2EI}{L}$	$-k_{AB}$ ⟶	$-C_{AB}k_{AB}$
$-\dfrac{2EI}{L}$ ⟵	$-\dfrac{4EI}{L}$	$-C_{BA}k_{BA}$ ⟵	$-k_{BA}$
$M = \dfrac{6EI}{L}$		$-k_{AB} - C_{BA}k_{BA}$	$-k_{BA} - C_{AB}k_{AB}$

Cantilever stiffness factor:

$M = EI/2$ Carry-over factor $= -1$

Fig. 7.5 – Various moment distribution coefficients.

In order to exploit these conditions three stiffnesses are described in Fig. 7.5. Anticipating symmetric problems such as the one above, using the symmetric stiffness factor implies that both ends of a member are rotated symmetrically and simultaneously. Similarly, the antisymmetric stiffness implies that both ends of a member are rotated antisymmetrically and simultaneously. The cantilever stiffness involves a unit rotation at one end (without lateral restraint) while the other end is fixed. It will be used in the solution below to allow the structure to translate as its joints are balanced thus avoiding the fictitious restraining force of Example 7.4. All of these stiffnesses may be computed directly using the beam equations or as simple moment distribution problems as indicated in the figure. The fixed end moments must either be computed or looked up in tables.

Since the symmetric case of Example 7.5 does not involve sidesway, it is trivial to solve using the symmetric stiffness for the beam and the regular stiffness for the column. Only one cycle of moment distribution is required.

The antisymmetric case is more interesting. If the antisymmetric stiffness used for the beam and the cantilever stiffness is used for the column, the structure moves laterally as the joint is balanced and no correction is required later for sidesway. The antisymmetric solution is then also trivial. The two moment diagrams are finally added for the solution of the original problem.

The symmetric and antisymmetric structures of Example 7.5 are themselves worthy of comment. In general, when a structure has a single axis of symmetry it is only necessary to deal with half of it when constructing a solution. But as indicated in this example, it is in general necessary to construct both an antisymmetric and a symmetric solution and superimpose the two. The fact that it is only necessary to deal with half the structure appears through the antisymmetric and symmetric structures which are each, of course, half the given structure with different boundary conditions in the two cases. The use of symmetry generally reduces the computational effort by about a factor of two and can thus be important in the analysis of large structures.

7.3.1 A Vierendeel 'Truss' Example

Example 7.6 shows a so-called Vierendeel 'truss' under what might be thought of as bridge loading. This type of structure is something of a

---

Example 7.6   Vierendeel 'truss'.

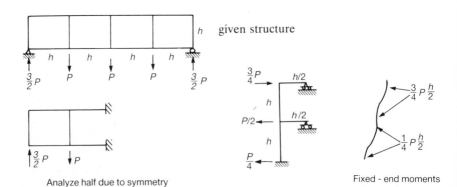

given structure

Analyze half due to symmetry

Fixed - end moments

Moment distribution: (coefficients of Ph)

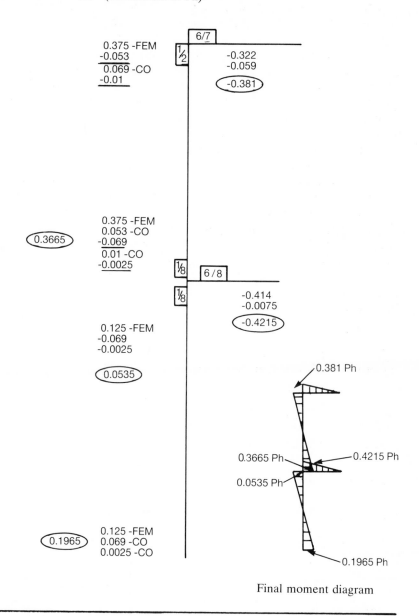

Final moment diagram

---

curiosity and not commonly found in practice except in cases where it is
demanded architecturally. It is, of course, not a truss (in the terms of this
text) but a rigid frame. Its behavior is quite different from that of a truss
since its loads are carried through bending. To make that point, it is

common to analyze such a structure approximately by assuming inflection points at mid-length of its members.

The analysis presented in Example 7.6 is interesting in its own right. In the first place the structure is symmetric so that it is only required to deal with half of it as indicated. Anticipating neglecting member length change, two fixed supports are used at center span. These supports would not be proper in the more general case in which length change is not neglected.

An additional use of symmetry is made in this example. When the half structure is rotated 90° it looks like a building under lateral load. Since the half frame is again symmetric about a vertical axis it is possible to use the symmetric/antisymmetric decomposition. Since the symmetric loading is zero it is finally only necessary to deal with the antisymmetric case which is similar to the antisymmetric case of Example 7.5. The final moment diagram is plotted.

A final word about the fixed-end moments in this case. Under lateral load with the ends of the columns fixed, the columns must have an infection point at mid-height. The fixed end moments are then equal to the story shear (the lateral force to be carried by the column) times half the story height.

7.3.2 Temperature and Settlement

Moment distribution is a convenient way to deal with certain classes of problems of temperature and settlement. For example, suppose the frame of Example 7.4 is subjected to a uniform temperature difference ΔT. It will deform symmetrically as indicated in Example 7.7 and can be handled easily using symmetric moment distribution stiffness factors. The fixed-end moments in this case come directly from Fig. 7.5.

Example 7.7   A frame under a uniform temperature increase ΔT.

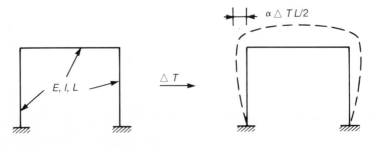

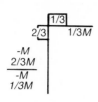

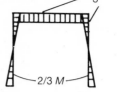

Moment distribution   Final moment diagram

Fixed-end moments $\dfrac{6EI\,\delta}{L^2} = \dfrac{6EI}{L^2}\,\dfrac{\alpha\,\Delta T L}{2} = \dfrac{3EI\,\alpha\,\Delta T}{L} = M$

Settlement problems require a similar type of fixed-end moment as indicated in Example 7.8.

Example 7.8   A settlement problem
Point A settles by an amount d

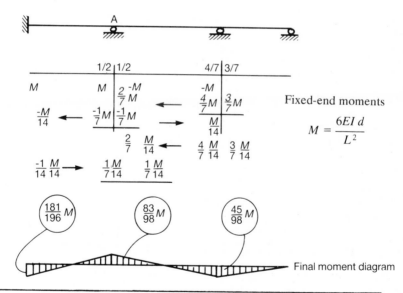

Fixed-end moments

$M = \dfrac{6EI\,d}{L^2}$

Final moment diagram

7.3.3. Variable Moment of Inertia
Example 7.9 is included here to make the point that there is a good deal of information available concerning members with variable moments of inertia. In this case the center member has constant moment of inertia while the end spans have linearly tapered haunches. The coefficients listed

Example 7.9 Variable moment of inertia problem.

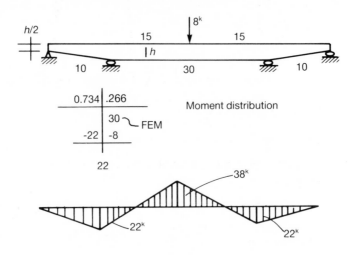

Fixed-end moments: $\dfrac{PL}{8} = \dfrac{8 \times 30}{8} = 30 \text{ K}'$

(Data from PCA *Handbood of Frame Constants*.)

$$C_{AB} = 0.834 \qquad C_{BA} = 0.294$$
$$k_{AB} = 6.86 \qquad k_{BA} = 19.46$$

Modified stiffness for roller support:

$$k'_{BA} = k_{BA}(1 - C_{AB}C_{BA})$$
$$= 19.46(1 - 0.834 \times 0.299) = 14.71$$

But PCA data are scaled to smaller end, therefore

$$k'_{BA} = 14.71 \frac{EI_{end}}{L} = 14.71 \frac{EI_{center}}{8L}$$
$$= 1.84 EI_{center}/L$$

Distribution factors:

$$\frac{1.84}{10} \quad \text{or} \quad 0.734, 0.266$$

$$\frac{2}{30} \quad \text{(symmetric distribution factor)}$$

in this example are taken from the PCA *Handbook of Frame Constants* and the example is relatively simple. The reader will find that problems of variable I requires some adjustment on his part but present no real difficulty.

7.4 EXERCISES

Solve the following problems using moment distribution. Use $E = 30 \times 10^6$ p.s.i., $\alpha = 0.65 \times 10^{-5}$ per °F.

Plot the moment diagrams.

1.
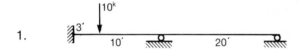

I – constant

2.
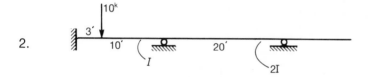

3.
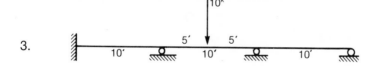

4.
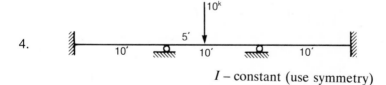

I – constant (use symmetry)

5.

I – constant (identify points of maximum moment)

6.

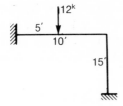

I – constant

7.

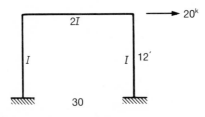

I – constant (identify points of maximum moment)

8.

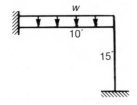

9.

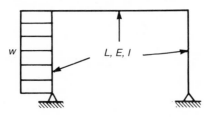

10.

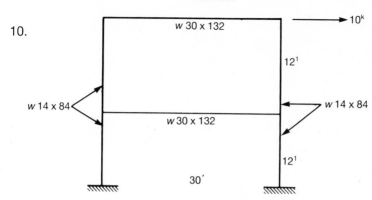

Problems **11–20**.   Resolve Problems **1–10** above for the case in which the load is zero and the left support settles 2″. Unless otherwise specified assume $I = 2000$ in⁴.

Problems **21–25**.   Resolve Problems **6–10** above for the case in which the load is zero and the structure is subjected to a uniform temperature change of 100 °F.

CHAPTER 8

Influence lines and their application

Fig. 8.1 shows a simply supported beam loaded by a unit force at some point x. It is desired in this case to compute the moment and shear at some point A. This is done in the figure by first computing the reactions in the usual manner then using free body diagrams which expose the moment and shear at point A. Finally, the moment and shear are plotted as a function of x keeping x_A fixed. These plots are called influence lines for moment and shear at point A.

In general, influence lines can be used to represent any type of solution in structures. They are like the Green's functions which arise in the study of differential equations in that they represent the effect of a concentrated unit force upon the particular quantity of interest. Influence lines would only be applied to linear problems since their utility lies in the idea of superposition. That is, the effect of a unit load, *per se*, is rarely of interest. The response to a unit load is only of interest as it can be scaled up to represent an arbitrary load or as it can be added to other solutions to represent the effect of several loads.

This chapter will only deal with influence lines for particular cases of stress resultants. In these cases, any influence line is concerned with two items. First, it is necessary to specify the type of loading to be used. (Certainly unit couples are as good as unit forces.) Second, it is necessary to specify the stress resultant of interest. (For example, the moment at point A or the shear at point A as discussed in Fig. 8.1.) Based on the linearity of classical structural analysis which allows solutions to be superimposed, three properties of influence lines are immediate:

1) For load of magnitude P, the stress resultant produced is P times the value of the influence line at the point of application of the load. (This follows from the fact that the influence line is constructed for a unit load.)
(2) The effect of a number of loads applied to a structure can be computed as the sum of the effects of single loads. (This is again superposition.)

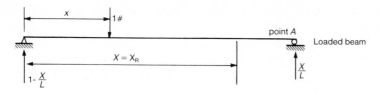

Free body diagrams

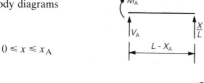

$0 \leq x \leq x_A$

$$V_A = -x/L$$

$$M_A = \frac{x}{L}(L - x_A)$$

$x_A \leq x \leq L$

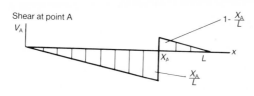

$$V_A = 1 - \frac{x}{L}$$

$$M_A = \left(1 - \frac{x}{L}\right)x_A$$

Plots of influence lines

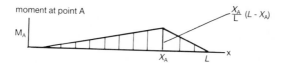

Fig. 8.1

(3) The effect of a uniformly distributed load can be computed as the area under the influence line times the magnitude of the distributed load. (This can be argued by approximating the distributed load by a number of concentrated loads and then allowing the number of loads to become very large.)

This chapter touches briefly upon many topics. The concept of an influence line is introduced as a simple and commonly used representation of a structural solution. Having introduced influence lines the beautiful Mueller–Breslau principle which is basic to structural engineering will be discussed. Finally some topics of bridge and building design will be introduced.

Before discussing these topics, Fig. 8.2 describes the construction of an influence line for the force in a bar of a truss. It should be noted that there is some ambiguity in the free body diagrams of Fig. 8.2 as the load

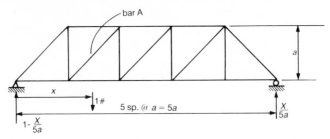

Free body diagrams

$0 \leqslant x \leqslant a$

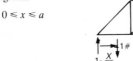

$$F_A = \sqrt{2}\,\frac{x}{5a}$$

$a \leqslant x \leqslant 5a$

$$F_A = -\sqrt{2}\left(1 - \frac{x}{5a}\right)$$

Influence line for the force in bar A

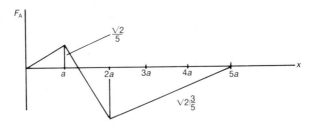

Fig. 8.2

moves between panel points of the truss. It is common in truss problems to construct points on influence lines for unit loads applied to joints and then later join these points with straight line segments. This is equivalent to assuming a statically determinate floor system which will transmit loads only to panel points.

8.1 THE MUELLER–BRESLAU PRINCIPLE

The Mueller–Breslau principle is basic to any discussion of influence lines and in many ways basic to any discussion of structures. In its most simple form it states that 'the influence line is the deflection curve'. Put more carefully, it should be stated as 'the influence line for any stress resultant is the deflection curve obtained by introducing in the structure a negative unit discontinuity corresponding to the stress resultant under discussion'. This is clearly true for the examples shown in Fig. 8.1. The influence line for moment has a unit discontinuity in slope at point A and the influence line for shear has a unit discontinuity in lateral displacement at point A. In fact, it holds for any elastic structure and can be quite useful, for example, in designing plates.

The Mueller–Breslau principle will only be proven here for the case of a single beam under lateral load such as the case shown in Fig. 8.3.

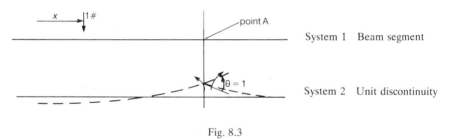

Fig. 8.3

Consider the virtual work of system 1 which represents a structure with a unit load at some point upon the displacements of system 2 in which there is a unit discontinuity at point A. The virtual work expressions developed earlier in Chapter 3 simply give in this case

$$1 \text{ lb} \cdot y_x + M \cdot \Delta\theta = 0 \Rightarrow M = -y_x$$

This argument can be generalized to any stress resultant and complex types of elastic structures.

8.2 AN EXAMPLE FROM BUILDING DESIGN

One of the most important features of the Mueller–Breslau principle is that it allows influence lines to be sketched without computation in some cases. This can provide important information concerning load placement. Consider, for example, the rigid frame of Fig. 8.4 which might represent a portion of a reinforced concrete building. It is usual to consider both 'dead load' (the weight of the structure) and 'live load' (people, equipment, . . .) in the design of structures of this type. There is, of course, no question concerning the placement of dead load since it must be placed wherever it occurs. Live load, on the other hand, must be placed in such a manner as to produce the worst possible effect.

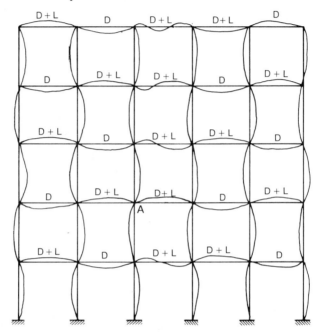

Rigid frame, sketch of influence line, and loading

Fig. 8.4 – Rigid frame, sketch of influence line, and loading

In Fig. 8.4 the influence line for negative moment at point A has been sketched assuming that sidesway is not an important factor under the given loading. This is done by guessing how the structure will respond to a unit discontinuity in slope at point A. Given this influence line it is necessary to decide how to load it in order to produce the worst possible effect. In this case it is clear that the positive areas of the influence line should receive the live load; loading the negative areas too would simply reduce the moment at A. What results is the so-called 'checkerboard' loading pattern in which in order to produce the maximum negative moment at point A, the two adjacent spans must be loaded plus alternate spans as you move away from point A. As you move up or down the spans to be loaded shift.

It is possible to carry this discussion one step further using what is called 'two-cycle moment distribution'. The argument is that in cases such as Fig. 8.4 under conditions of live and dead load, what is done away from the point of interest has little effect upon this point. In order to find the maximum dead plus live load moment at point A then it is only necessary to consider joints which are immediately adjacent to it. (See Fig. 8.5.) Using two-cycle moment distribution this is done by starting with joint A

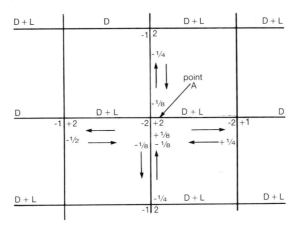

Assume: Final moment at A
(1) All dist. factors $\frac{1}{4}$ 2
(2) Dead load = Live load   + 1/8
                                 + 1/16
                                 2 3/16

Fig. 8.5

and balancing it, carry-over is then made to affected adjacent joints which are themselves then balanced, finally carry-over in made back to joint A which is then balanced.

Two-cycle moment distribution can be an effective tool in the design of reinforced concrete buildings.

8.3 MOVING LOADS

In building design it is most common to either apply live load to an entire span or leave the span unloaded; partial loading is not common. In bridge design, on the other hand, there is a concern with where to place trucks to produce the worst possible effects. An example of this is discussed in Fig. 8.6 which is concerned with the question of the maximum live load shear which can occur at any point in a simply supported span. It would be required to know this in order to design shear connectors in composite construction or cover plate details in a plate girder.

The starting point of this discussion is the influence line for shear which was developed in Fig. 8.1. The three loads to be used which are shown in Fig. 8.6 comprise an AASHTO truck – a common type of bridge design load. For this truck and this application the load spacing must be kept fixed but the 'load group' may be placed anywhere on the span. It is obvious from the shape of the influence line that the maximum shear will be produced when the largest axle load is placed at the peak value of the

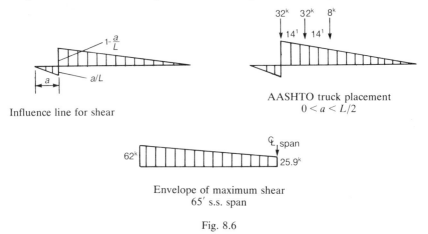

Influence line for shear

AASHTO truck placement
$0 < a < L/2$

Envelope of maximum shear
65' s.s. span

Fig. 8.6

influence line with the other loads coming out where they must. For a 65' span the maximum shear at any point 'a' $(0 \leq a \leq L/2)$ can be computed by summing the effects of the three truck axle loads or by taking the resultant load (72^K, see Fig. 8.7) times its influence line ordinate.

$$V_{max} = 72(65 - a - 9.33)/65 = 72(55.67 - a)/65$$

When this is plotted for point 'a' ranging over half the span the 'envelop of maximum shear' for half the span is obtained.

8.4 PLACING MOVING LOADS FOR MAXIMUM MOMENT

Even when an influence line is available, there are situations in which it is not clear how the loads should be placed on the span to produce a maximum effect. In this section a simple result is developed which can be useful in determining how loads should be located in order to produce maximum bending moment.

Fig. 8.6 shows a simply supported span over which a set of loads such as those which are used in railroad or highway bridge design specifications is allowed to move. It would be useful in this case to have a result which would indicate where to place the load (actually the load group) in order to produce the maximum possibile moment in the span, but unfortunately no such result is available. Here it will only be possible to select one of the concentrated loads and find the location of the load group so that the moment under this load is maximum. It will be shown that for maximum

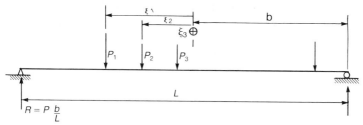

Moving load group

For maximum moment under load P_1 set $b = L - \xi_1 - b$

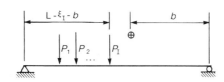

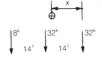

Centroid: $\bar{x} = \dfrac{8 \times 28 + 32 \times 14}{72}$

$= 9.33'$

AASHTO truck

Maximum moment under center load.

$$D = \frac{65 - 4.67}{2} = 30.16'$$

$M_1 = 33.41 \times 30.16 - 8 \times 14 = 895.6 \text{ K}'$

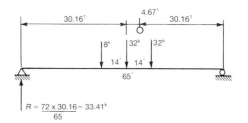

effect the load group should be located so that the distance from the centroid of the load group from the right support equals the distance from the left support to the load under consideration.

In this figure there are n loads which are allowed to move over the span while the distances between these loads remain fixed. This load group has a centroid as indicated in the figure and the known quantities ξ_i locate the individual loads with respect to this centroid. Let

$$P = \sum_{i=1}^{n} P_i$$

represent the total applied load. The left reaction can then be computed as

$$R = \frac{Pb}{L}$$

and a free body diagram used to compute the moment M_I under the Ith load as

$$M_I = \frac{Pb}{L}(L - \xi_I - b) - P_1(\xi_1 - \xi_I) - P_2(\xi_2 - \xi_I) - \cdots$$

It is now possible to maximize M_I with respect to the term b as

$$\frac{dM_I}{db} = P - \frac{2Pb}{L} - \frac{P\xi_I}{L} = 0 \Rightarrow b = L - b - \xi_I$$

This is the result stated above.

Fig. 8.6 shows how this result can be applied to a typical AASHTO truck on a 65′ span.

8.5 EXERCISES

1 Compute and plot the influence line for the horizontal reaction of a symmetric three-hinged parabolic arch.

2 Compute and plot influence lines for members a, b, c, d, e in the given truss.

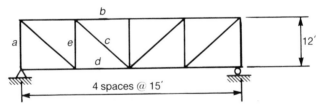

3 Use the building of Fig. 8.4 and its loading and compute the maximum center span moment (in the span of point A) using two-cycle moment distribution.

4 The AASHTO specifications state that for a 100′ span
(a) the maximum design bending moment is 1524 K′
(b) the maximum shear is 65.3^K.
Use the truck shown in Fig. 8.7 and show that these values are correct.

References

CHAPTER 1

There are any number of excellent mechanics texts (statics and dynamics) which emphasize two-dimentional applications. For three-dimensional rigid body mechanics the choice is somewhat more restricted. In this case the reader may wish to consult

Herbert Goldstein, *Classical Mechanics*, Addison-Wesley, Reading, Mass., 1950.
John L. Synge and Byron A. Griffith, *Principles of Mechanics*, McGraw-Hill Book Co, New York City, 1949.

CHAPTER 2

For the topic of linear algebra the standard text is commonly regarded to be

Ben Noble, *Applied Linear Algebra*, Printice-Hall, Englewood Cliffs, NJ, 1969.

An abbreviated but highly readable version of some of the same material can be found in

A. C. Aitken, *Determinants and Matrices*, Oliver and Boyd, London, 1958.

There are many excellent texts on structures on the market today. Two of the older classics are

Charles Head Norris, John Benson Wilbur, and Senol Utku, *Elementary Structural Analysis*, McGraw-Hill Book Co, New York City, 1976.
Hale Sutherland and Harry Lake Bowman, *Structural Theory*, John Wiley and Sons, New York City, 1950.

The latter book has a good account of the Wichert truss and references to this topic this topic by the famous American bridge engineer D. B. Steinman.
In the area of matrix analysis of structures there are many excellent texts available. For obvious reasons the author wishes to reference

William R. Spillers, *Automated Structural Analysis: an Introduction*, Pergamon Press, New York, 1972.

A new and very readable matrix structural analysis text is

T. R. Graves Smith, *Linear Analysis of Frameworks*, Ellis Horwood Publishers, Chichester, 1983.

There is surprisingly little material on curved beams available in texts and technical papers. In any case the reader might wish to look at the comprehensive study of:

Eric Reissner, 'Variational Considerations for Elastic Beams and Shells', Journal of the Engineering Mechanics Division, Proceedings American Society of Civil Engineers, 88 EM1 Feb 1962, pp. 23–57.

The common American reference text in shell theory is

S. Timoshenko, *Theory of Plates and Shells*, McGraw-Hill Book Co, New York City, 1940.

For membrane shell theory there is the very readable book

Alf Pfluger, *Elementary Statics of Shells*, F. W. Dodge Corp, New York City, 1961.

CHAPTER 3

A general discussion of the application of the method of virtual work to problems of structures is available in

George A. Hool and W. S. Kinne, *Movable and Long Span Steel Bridges*, McGraw-Hill Book Co, New York City, 1943.

CHAPTER 5

Two of the standard texts on plastic analysis are

Philip Hodge, Jr., *Plastic Analysis of Structures*, McGraw-Hill Book Co, New York City, 1959.
B. G. Neal, *The Plastic Methods of Structural Analysis*, John Wiley and Sons, New York City, 1956.

For a comprehensive discussion of plastic analysis applied to three-dimensional frames see

C. A. Gonzales and S. J. Fenves, 'A Network-Topological Formulation of the Analysis and Design of Rigid Plastic Framed Structures', Report no. 339, Department of Civil Engineering, Univ of Illinois, Urbana, Sept 1968.

Available steel structural members are described in the

Manual of Steel Construction, American Institute of Steel Construction, Chicago, 1980.

In this same vein, there is what can only be regarded to be a gem for structural engineering students

R. L. Brockenbrough and B. G. Johnston, *USS Steel Design Manual*, United States Steel Corporation, 1968.

This small volume comes close to containing everything the structural engineer needs to know about steel structures.

CHAPTER 7

The most complrehensive text available on the subject of moment distribution is

J. M. Gere, *Moment Distribution*, D. van Nostrand Co, New York City, 1963.

An excellent source of moment distribution coefficients for problems of variable moment of inertia is available in

Handbook of Frame Constants, Portland Cement Association, Skokie, Illinois 60077, 1958.

APPENDIX 3

Integral transforms provide an excellent way of dealing with discontinuous functions. See, for example,

Balth van der Pol and H. Bremmer, *Operational Calculus Based on the Two-Sided Laplace Integral*, Cambridge University Press, Cambridge, 1964.

APPENDIX 4

The classic text in the area of differential geometry is

Dirk J. Struik, *Differential Geometry*, Addison-Wesley, Reading, Mass., 1961.

APPENDIX 6

One of the few comprehensive discussions of the topic of statical indeterminancy has been presented by

Frank DiMaggio, 'Statical Indeterminancy and Stability of Structures', Journal of the Structural Division, American Society of Civil Engineers, June 1963, pp. 63–75.

APPENDIX 1*

The rotation matrix

A1.1 PLANE ROTATION

When a vector is viewed in two coordinate systems which differ only by a rotation, there is a geometric relationship between the components in the

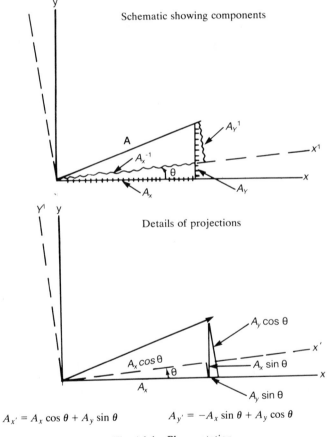

$$A_{x'} = A_x \cos \theta + A_y \sin \theta \qquad A_{y'} = -A_x \sin \theta + A_y \cos \theta$$

Fig. A1.1 – Plane rotation

two systems which is indicated in Fig. A1.1. The scalar relationship of that figure can be written compactly in matrix notation as

$$\begin{bmatrix} A_{x'} \\ A_{y'} \end{bmatrix} \begin{bmatrix} \cos\theta & \sin\theta \\ -\sin\theta & \cos\theta \end{bmatrix} \begin{bmatrix} A_x \\ A_y \end{bmatrix} \tag{A1.1}$$

or symbolically as

$$A' = RA \tag{A1.2}$$

Here R is called the two-dimensional rotation matrix.

A1.2 THREE-DIMENSIONAL ROTATIONS

A1.2.1 Compound Rotations
One way of dealing with a three-dimensional rotation is to describe it is a sequence of rotations about coordinate axes. Since rotations about coordinate axes are essentially plane rotations, this approach has the effect of reducing general three-dimensional rotations to sequences of plane rotations.

For rotations about coordinate axes the rotation matrices are

$$R_x = \begin{bmatrix} 1 & 0 & 0 \\ 0 & \cos c & \sin c \\ 0 & -\sin c & \cos c \end{bmatrix} \quad \text{(rotation of } c \text{ degrees about the } x \text{ axis)}$$

$$R_y = \begin{bmatrix} \cos b & 0 & -\sin b \\ 0 & 1 & 0 \\ \sin b & 0 & \cos b \end{bmatrix} \quad \text{(rotation of } b \text{ degrees about the } y\text{-axis)}$$

$$R_z = \begin{bmatrix} \cos a & \sin a & 0 \\ -\sin a & \cos a & 0 \\ 0 & 0 & 1 \end{bmatrix} \quad \text{(rotation of } a \text{ degrees about the } z\text{-axis)}$$

Here the matrix R_z follows trivially from Eq. (A1.1) since a plane rotation can be thought of as a rotation about the z-axis which comes out of the paper; the other two matrices follow directly by permuting the xyz-axes carefully.

The idea of a compound rotation matrix comes from the fact that a sequence of rotations is described by taking the *product* of the associated matrices (in the proper order). For example, a sequence of rotations producing

$$x \rightarrow x' = Rx$$
$$x' \rightarrow x'' = R'x'$$
$$x'' \rightarrow x''' = R''x''$$

results in a compound composite rotation matrix as

$$x''' = R''x'' = R''R'x' = R''R'Rx$$

The rotation matrix from x to x''' appears as a product as suggested above,

$$x''' = R^*x \Rightarrow R^* = R''R'R$$

When R'', R', R are each rotations about coordinate axes, they can be taken directly from Eq. (A1.3).

A1.2.2 Rotation Matrices Constructed from Base Vectors

Another method of constructing a three-dimensional rotation matrix is based on

THEOREM A.1.1. *The rows of a rotation matrix are composed of the base vectors (coordinate unit vectors) of the local coordinate system.*

First, note that the original or unprimed coordinate system will be called the global coordinate system and the rotated or primed system will be called the local coordinate system. In order to demonstrate the above theorem it is convenient to use another theorem.

THEOMEN A1.2. *The transpose of a rotation matrix is equal to its inverse:* $\tilde{R} = R^{-1}$.

Proof. Under a rotation $A' = RA$, by definition, the magnitude of a vector does not change. Therefore

$$|A|^2 = \tilde{A}A = \tilde{A}'A' = \tilde{A}\tilde{R}RA$$

or

$$\tilde{A}(I - \tilde{R}R)A = 0$$

which implies that $\tilde{R}R = I$.

If $\tilde{R} = R^{-1}$, the inverse transformation follows directly as

$$A' = RA \Rightarrow A = \tilde{R}A'$$

Now let

$$A' = \begin{bmatrix} 1 \\ 0 \\ 0 \end{bmatrix} = \mathbf{i}',$$

the unit vector in the x' (the local x) direction. By direct multiplication it follows that in this case

$$A = \tilde{R}A' = \begin{bmatrix} R_{11} \\ R_{12} \\ R_{13} \end{bmatrix} = \begin{bmatrix} (\mathbf{i}')_x \\ (\mathbf{i}')_y \\ (\mathbf{i}')_z \end{bmatrix}$$

That is, the first row of the rotation matrix is a unit vector in the x' or local x direction. Similar results follow for the other rows. Finally,

$$R = \begin{bmatrix} (\mathbf{i}')_x & (\mathbf{i}')_y & (\mathbf{i}')_z \\ (\mathbf{j}')_x & (\mathbf{j}')_y & (\mathbf{j}')_z \\ (\mathbf{k}')_x & (\mathbf{k}')_y & (\mathbf{k}')_z \end{bmatrix} \tag{A1.4}$$

or symbolically

$$R = \begin{bmatrix} \mathbf{i}' \\ \mathbf{j}' \\ \mathbf{k}' \end{bmatrix} \tag{A1.5}$$

APPENDIX 2

Force–displacement duality of geometric instability

There is a theorem of linear algebra which states that either

$$Ax = b \text{ has a solution } x \neq 0 \text{ for arbitrary } b$$

or

$$\grave{A}y = 0 \text{ has a solution } y \neq 0.$$

Proof:
(1) Assume that there exists an $x \neq 0 \mid Ax = b$. Multiply by $y \neq 0$: $\bar{y}Ax = \bar{y}b \neq 0$. However, $\bar{y}b$ would have to be zero if $\grave{A}y = 0$ were also valid.
(2) Assume that there exists a $y \neq 0 \mid \grave{A}y = 0$. Multiply by $x \neq 0 : \bar{x}\grave{A}y = 0$. If x were also to satisfy $Ax = b$ it would imply that $\bar{x}\grave{A}y \neq 0$.

The structural analog of this theorem states that either:

(1) The equilibrium equations have a solution for arbitrary load
or
(2) There exists a rigid body motion (i.e. a set of non-zero node displacements which produce no member displacements).

Discontinuous functions

It has been argued in Chapter 2 that it is sufficient to limit discussions of the differential equations for beams to the case of distributed loads (and torques) since point or concentrated loads can be handled as a special case of distributed loads. This appendix discusses how to do so using discontinuous functions.

Fig. A3.1 shows schematically what is involved. Consider, for

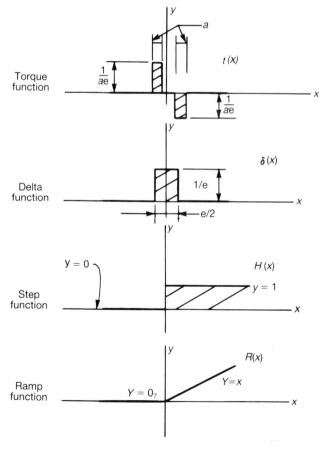

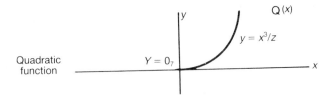

Quadratic
function

Note that these functions are related through their derivatives:

$$\frac{d}{dx}\,\delta = t$$

$$\frac{d}{dx}\,H = \delta$$

$$\frac{d}{dx}\,R = H$$

$$\frac{d}{dx}\,Q = R$$

Fig. A3.1 – A sequence of discontinuous functions

example, the Dirac delta function $\delta(x)$ indicated there which corresponds to a unit point load. The idea is that as $\varepsilon \to 0$ the area under the function stays constant at unity. Note that this would be required of any definition of a point load as some limiting case of a distributed load. Similarly, the derivative of the delta function can be argued to correspond to a unit concentrated torque. Operationally, these discontinuous functions require two assumptions:

(1) It is assumed that the mean value theorem of calculus is valid in the sense that

$$\int_a^b f(x)\,\delta(x-c)\,dx = \begin{cases} f(c) & a < c < b \\ 0 & \text{otherwise} \end{cases} \tag{A3.1}$$

where $f(x)$ is any well-behaved function.

(2) It is assumed that integration by parts is valid for these discontinuous functions $d(x)$, i.e.

$$\int_a^b f(x)\,d'(x)\,dx = fd\,\Big|_a^b - \int_a^b f'd\,dx$$

or

$$\int_a^b f'(x)\,d(x)\,dx = fd\,\Big|_a^b - \int_a^b fd'\,dx \tag{A3.2}$$

Note finally that the functions of Fig. A3.1 are related and that as you move up in the figure each function is the derivative of the one below it.

Example A3.1   Find the shear and moment.

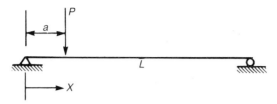

Integrate the differential equation

$$\frac{d^2M}{dx^2} = -w \qquad \text{when } w = \delta(x - a) \,.\, \mathbf{P}$$

$$\frac{d^2}{dx^2} M = -P \,.\, \delta(x - a)$$

$$V = \frac{d}{dx} M = -PH(x - a) + c_1$$
$$M = -PR(x - a) + c_1 x + c_2$$

Boundary conditions:

$$M = 0 \ @ \ x = 0 \Rightarrow c_2 = 0$$
$$M = 0 \ @ \ x = L \Rightarrow -P(L - a) + c_1 L = 0$$

$$c_1 = P \frac{L - a}{L}$$

$$V = P\left(-H(x - a) + \frac{L - a}{L}\right)$$

$$M = P\left(-R(x - a) + x \frac{L - a}{L}\right)$$

Diagrams:

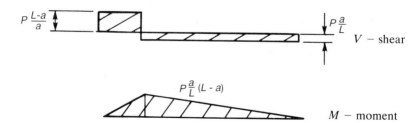

$$P\frac{L-a}{a}$$

$$P\frac{a}{L}$$

$$V - \text{shear}$$

$$P\frac{a}{L}(L - a)$$

$$M - \text{moment}$$

The Frenet Formulas

In deriving the differential equations of equilibrium for curved beams in Chapter 2, expressions were required for the derivatives of unit vectors. They are called the Frenet formulas:

$$\mathbf{t}' = \mathbf{n}/\rho$$
$$\mathbf{n}' = \mathbf{b}/\tau - \mathbf{t}/\rho \qquad\qquad (A4.1)$$
$$\mathbf{b}' = -\mathbf{n}/\tau$$

Their derivation is discussed briefly in this appendix.

The starting point for this discussion is a curve in space represented by the position vector $\mathbf{r}(s)$ given as a function of the arc length s. (See Fig. A4.1.) The first derivative of $\mathbf{r}$ with respect to s clearly defines the tangent unit vector to the curve $\mathbf{t}$

$$\frac{d}{ds}\mathbf{r} = \mathbf{r}' \equiv \mathbf{t} = \lim_{\Delta s \to 0} \frac{\Delta \mathbf{r}}{\Delta s} \qquad\qquad (A4.2)$$

The second derivative of $\mathbf{r}$ with respect to s defines the normal direction but turns out not to be a unit vector, i.e.

$$\mathbf{r}'' = \mathbf{t}' = \mathbf{n}/\rho \qquad\qquad (A4.3)$$

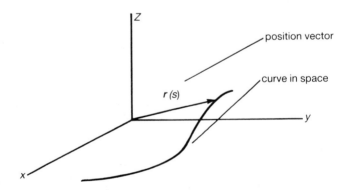

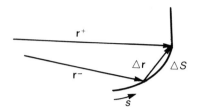

The tangent vector as a derivative

$$r' = \frac{dr}{ds} = t = \lim_{\Delta s \to 0} \frac{\Delta r}{\Delta s}$$

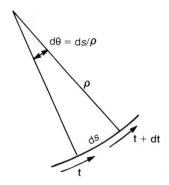

The radius of curvature ρ

$$\left| \frac{dt}{ds} \right| = \frac{1}{\rho}$$

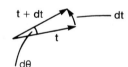

$$\psi v \Sigma \gamma \delta \; |t| = 1 \qquad |dt| = d\theta$$

Fig. A4.1 – The Frenet formulas

where ρ is called the *radius* of the curve and **n** is a unit vector. Note first that $\mathbf{n} \perp \mathbf{t}$ since

$$\mathbf{t} \cdot \mathbf{t} = 1 \Rightarrow \mathbf{t}' \cdot \mathbf{t} = 0 \tag{A4.4}$$

The designation of ρ as a radius of curvature follows from the fact (see Fig. A4.1) again that

$$|\mathbf{dt}| = |d\theta| \cdot 1 = |ds/\rho| \cdot 1 \Rightarrow \left| \frac{dt}{ds} \right| = \frac{1}{\rho} \tag{A4.5}$$

Given **t** and **n** the binormal direction **b** is defined as

$$\mathbf{b} = \mathbf{t} \times \mathbf{n}$$

At this point the position vector $\mathbf{r}(s)$ has been used to construct an *intrinsic* coordinate system with base vectors $\mathbf{t}$, $\mathbf{n}$, $\mathbf{b}$ at every point of the curve. Since $\mathbf{n}$ is a unit vector, $\mathbf{n}' \perp \mathbf{n}$ (see Eq. (A4.4) again) and can be written in terms of the constants A and B as

$$\mathbf{n}' = A\mathbf{t} + B\mathbf{b} \tag{A4.6}$$

but

$$\mathbf{t} \cdot \mathbf{n} = 0 \Rightarrow \mathbf{t}' \cdot \mathbf{n} + \mathbf{t} \cdot \mathbf{n}' = 0 \tag{A4.7}$$

or

$$\mathbf{t} \cdot \mathbf{n}' = A = -\mathbf{t}' \cdot \mathbf{n} = \frac{-\mathbf{n}}{\rho} \cdot \mathbf{n} = -\frac{1}{\rho} \tag{A4.8}$$

The term B is simply defined to be one divided by the *torsion* τ.

Again $\mathbf{b}' \perp \mathbf{b}$ which implies that

$$\mathbf{b}' = C\mathbf{t} + D\mathbf{n} \tag{A4.9}$$

but

$$\mathbf{t} \cdot \mathbf{b} = 0 \Rightarrow \mathbf{t}' \cdot \mathbf{b} + \mathbf{t} \cdot \mathbf{b}' = 0$$

or

$$\mathbf{t} \cdot \mathbf{b}' = C = -\mathbf{t}' \cdot \mathbf{b} = \frac{-\mathbf{n}}{\rho} \cdot \mathbf{b} = 0 \tag{A4.10}$$

Finally,

$$\mathbf{n} \cdot \mathbf{b}' = D = -\mathbf{n}' \cdot \mathbf{b} = -\mathbf{b} \cdot (\mathbf{b}/\tau - \mathbf{t}/\rho) = -\frac{1}{\tau} \tag{A4.11}$$

This completes the derivation of Eq. (A4.1).

The use of the term torsion has physical significance. If $\mathbf{t}$ and $\mathbf{n}$ are thought to define a plane at any point on the curve, the binormal vector $\mathbf{b}$ is a unit normal to this plane. Following the earlier discussion of the unit tangent $\mathbf{t}$, and term $d\mathbf{b}$ represents an angle change $d\phi$ which is, in fact, the twist (torsion) of the vector $\mathbf{t}$. The associated scalar τ then has the appearance of a 'radius of curvature' if

$$d\phi \equiv ds/\tau \tag{A4.12}$$

structure. When it has not

$$k = \begin{cases} 3m - r & \text{plane frames} \\ 6m - r & \text{space frames} \end{cases}$$

Here r is the number of releases.

This procedure has been carried out in Fig. A6.2 for a truss discussed in Chapter 4. In a rather cumbersome manner, hinges are added to a plane frame to create the desired truss. For small structures and practical purposes it is adequate to select redundants by trial and error; for the truss under discussion this means cutting bars until a statically determinate and stable structure is achieved. Note that while k can be shown to be invariant when a structure has been given, the selection of redundants can lead to different reduced structures as indicated in Fig. A6.2.

APPENDIX 7

Some beam solutions

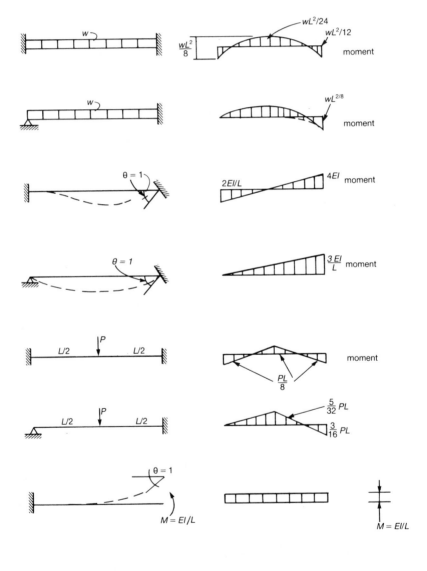

A transcendental equation

The transcendental equation

$$\cosh(a + b) - \cosh(b) = c \qquad (A8.1)$$

can be solved for b given a and c in the following manner. First, there is an addition formula for the cosh which states that

$$\cosh(x + y) - \cosh(x - y) = 2 \sinh x \sinh y \qquad (A8.2)$$

Now let

$$x = \frac{a}{2} + b \quad \text{and} \quad y = a/2$$

Then Eq. (A8.1) can be written as

$$\cosh\left(\frac{a}{2} + b + \frac{a}{2}\right) - \cosh\left(\frac{a}{2} + b - \frac{a}{2}\right) = 2 \sinh\left(\frac{a}{2} + b\right)\sinh\left(\frac{a}{2}\right) = c$$

or

$$\frac{a}{2} + b = \sinh^{-1}\left[\frac{c}{2 \sinh(a/2)}\right]$$

and

$$b = \sinh^{-1}\left[\frac{c}{2 \sinh(a/2)}\right] - a/2$$

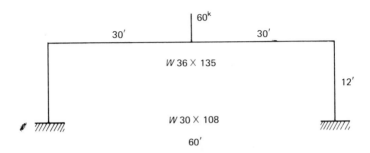

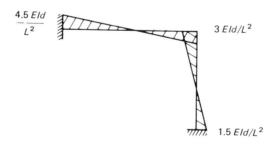

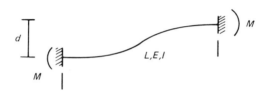

Index